GREAT CARS

SPORTS AND RACING

Doug Nye

Galley Press

Author Doug Nye
Illustrator Jim Dugdale
Editor Tim Auger
Designer Ron Pickless
Consultant Michael Ware

Published in this edition by
Galley Press, an imprint of
W. H. Smith and Son Limited,
Registered No. 237811 England.
Trading as WHS Distributors,
St John's House, East Street,
Leicester, LE1 6NE.

ISBN 0 86136 829 0

Printed and bound in Portugal by Printer Portuguesa, Sintra.

Contents

FRANCE
4
elf
ALPINE
935 HB 76

SPORTS CARS

Introduction

What is a sports car? That question has been asked innumerable times, and many long and learned essays have been written under that title. One could claim that a sports car is an open two-seater offering high performance both in a straight line, and through corners. But some sports cars have appeared which have roofs and more than two seats and which can out-perform their open roadster brethren not only along the straight, but also through those all-important corners. So we try another answer. A sports car is a high-performance car. Simple as that? Well, no, for some high-performance cars are built very much for straight-line comfort; to carry five people and a mass of luggage a thousand miles or more on motorways, freeways, autoroutes and autostrada with a minimum of fuss and fatigue. A Citröen-Maserati SM saloon car or a Rolls-Royce Corniche is a high-performance car of this type. With the best will in the world, no enthusiast could adequately describe either of them as a 'sports car'.

Perhaps a sports car is largely indefinable, but those who seek one will recognize it when they see it . . .

For many years the type in essence was virtually unmistakable. Most sports cars would have open cockpits, and most of them would have no more than two seats. The type evolved in the vintage period of the nineteen-twenties and caught firm hold in the thirties in models such as the MG Midget and Magnette. Postwar their descendants, particularly the MG T-series, spelled 'sports car' to the world in an unmistakable manner. Yet the T-series cars in particular, the MG TC, TD and TF of the late forties and early fifties, were never particularly fast cars. What was really important was that they looked fast, they looked sporty, and the people who owned and drove them with such enthusiasm (despite the miseries of leaking hoods and freezing feet in rain and snow) derived immense *enjoyment* from them. This is perhaps the key. For the sake of enjoyment, for many years men and women took up open-air motoring in a machine which would accelerate more briskly than most bread-and-butter 'transport' cars on the road; which would brake more effectively than them; which might have a higher ultimate top speed in a straight line; and which would especially be capable of negotiating corners safely at a much higher speed. These attributes in turn produced a vehicle admirably well-suited for use in competition events on road or track in addition to humdrum motoring duties. If a car was essentially capable of participating in sporting events, then it could be assumed to be a 'sports car'.

One must have reservations about such a definition, but looking through the following pages the reader will see that the sports cars we have selected do not stray far on either side of a sporting central thread. The earliest 1903 60 hp Mercedes was a high-performance production road car stripped to save weight and entered as a last minute stop-gap in the important Gordon Bennett Trophy race, which it actually won. In this way it proved itself a sporting car of considerable stature, although in essence its design was derived from that of a pure-blooded racer, 'de-tuned' for road use. Here the sports car shows an origin close to that of the Grand Prix car, but this book follows the sports car as it developed along separate lines.

The first true sports-car competition developed in Germany before World War I as the Herkomer Trial, and its successful descendant, the Prince Henry Trial. These events almost unwittingly encouraged interested motor-manufacturers to construct specialized touring cars which combined powerful engines with lightweight body designs and the latest thinking in steering, braking and suspension. Upon the success of the Trial-winning designs, victorious manufacturers would put into small-series production cars which were in many cases sold only to customers of recognized ability! In many cases these carefully allocated sporting cars were destined to undertake extensive sprint and hillclimb competition programmes in order to promote their manufacturers' names and images more widely, and upon their success depended the mass sales of those 'transport' designs–the growing volume of bread-and-butter motor cars designed to transport people and baggage from point A to point B . . . and little more.

After World War I came the vintage years of motoring in Europe and America. This was the period in which the many-faceted 'sports car' market developed, although not under such a glib and modern title. Enthusiast cars ranged from the spartan and often diabolically dangerous cyclecars like the various GNs of Britain and the French Bedélia designs to the middle-of-the-road Alvises, the strikingly neat and fast Amilcars, Salmsons and Bugattis, the dignified but so sophisticated Lancias, to the sporting epitome of their time–the likes of the Vauxhall 30/98, the Bentleys and the SS-series Mercedes-Benz.

Today each of these cars, even the still-humble but so fascinating little cyclecars, have become much-sought-after collectors' items, cherished by the true enthusiast and salted away by the stone-hearted investor. What most of these classical vintage sports cars have in common is their lack of a fixed roof and their considerable power/braking/steering/roadholding advantage over contemporary workaday saloon cars. This advantage in these vital departments continued into the thirties although some exceptions, like the saloon-bodied Lancias and the mighty Hispano-Suiza limousines, served to prove the rule. Despite their extra weight both were capable of out-performing many more flashy 'true sports cars' upon otherwise even terms.

During the thirties sports-car racing achieved major importance in events such as the Le Mans 24-Hours in France, the Mille Miglia in Italy and the RAC Tourist Trophy in the United Kingdom. These fostered the design, construction and promotion of such classic sports cars as the Alfa Romeo 8C-2300, the MG 'K3' Magnette, the Talbot 105 and the Delahaye 135C. Each of these great cars is described in detail in the text which follows, and is fixed there in relation to its

The 1937 works team of SS '100' models posed before the year's Welsh Rally.

◀ *Bugatti Type 55 hill-climbing in Britain, inside rear wheel spinning, tyre smoking, and no doubt the exhaust note setting every spectator's spine a-tingle. . . .*

▼ *1925 Lancia Lambda storming the stony hill at Beggar's Roost during a 'tween-wars Lands End Trial.*

sister models from the same manufacturer, and against the background of its contemporary rivals.

The vital thing to remember is that competition models such as these were developed from, and became the prototypes for further development of high-performance sporting cars intended primarily for sale to the general public for use on the open highway. In this vital respect they differed from the highly-specialized purpose-built racing car, which had developed worldwide into a single central-seat open-wheeled projectile by 1932–33 and which in essence was to have decreasing significance for road-car manufacturers as the years rolled by.

Yet in the thirties the sporting car was still essentially a wealthy man's toy, and there is no doubt that the lucky original owners of these cars often played long and hard with them, and derived enormous pleasure from the experience. It was after World War II that such pleasure really began to become available to all, in the mushroom growth particularly of the British sports car industry in which models like the immortal MG T-series, the Triumph TRs, the Austin-Healey family, the mighty Jaguar range and the peripheral marques like the Allards,

HRGs, the true Healeys, Aston-Martins were produced *en masse*. The vast majority of all these British cars were exported, mainly to the USA, and as Britain came to displace Italy from motor-racing domination so this firm foundation of sports-car know-how and commercial success played its part.

Still, the Italian sports car had made its mark through the 'tween-war years and that turbulent nation of flair, imagination and supreme craftsmanship continued to produce sports cars which represent the pinnacle of the art – names like Ferrari, Maserati and Lancia were to have few peers in international road-racing competition, but their products were few and expensive, beyond the pocket of Mr Average, who could easily find the wherewithall to indulge his enthusiasm at the wheel of an MG, TR, Healey or 'Jag'.

The success of these marques in all parts of the world was not lost upon General Motors and the Ford Motor Company in America, and in the fifties and sixties both launched their sports car projects – the Chevrolet Corvette family and the Mustang-to-GT40 programme which in both cases had a two-fold effect. Firstly sporting cars were sold to a new-found 'youth' market with the money to pay for fun on four wheels. Secondly these sporting cars in general image and in all manner of competition – from the most humble Sports Car Club of America club race to the International arena of Le Mans – promoted a feeling of endeavour, excitement and achievement which reflected credit throughout the rest of those manufacturers' ranges, however mundane.

By this time the sports car as generally accepted – the open-top two-seat roadster – had begun to lose its performance advantage over the bread-and-butter transport motor car. More competition sporting-car know-how was being built into more cars. Ford of Britain achieved saloon-car racing dominance in Britain and her Commonwealth with successive models from the 105E Anglia through the Cortina to the Escort and Fiesta of today. Alfa Romeo in Italy picked up with the GTA saloon where they had once dominated with the classical 8C-2300; the class of racing had changed, but the success and image of achievement was the same. Jaguar saloon cars for many years were just as dominant in British saloon-car racing as had been their C-Type and D-Type sports-racing sisters in former years at Le Mans. Mercedes-Benz sent saloon cars racing successfully through the Pampas of Argentina and round the looping mountain course of the Nürburgring to emulate the former dominance of their legendary Grand Prix cars, and of the immortal 300SLR sports-racer of the fifties.

Rallying Alpine-Renault.

In each case these were tin-top saloon cars which in standard trim could be seen flowing through (or choking) the streets of any town or city almost anywhere in the Free World. The sixties saw the release of saloon car models like the Ford Lotus-Cortina and the Escort Twin-Cam, which could out-brake and often out-corner many of the classic sports cars of yore, and most of their contemporary pure-bred 'sports car' competition. For example any reasonable Ford Escort saloon would make most MGs or Triumph TRs look positively lame through winding country lanes, and so a form of roofed-in but still pleasurable sporting motoring evolved. This reached its natural conclusion in the development of enclosed-cockpit sports cars, such as the original Triumph TR7, which was hastily designed when it appeared that the burgeoning volume of American Federal Safety Regulations would ban open-cockpit cars, because of their lack of roll-over protection for their occupants. This subsequently proved to be nothing more than scare-mongering, and an open-cockpit TR7 has now been introduced specifically for the lucrative but sensitive American market.

Towards the end of this book, therefore, you will find many roofed-in high-performance cars which qualify for the 'sports car' tag none the less. The Ford Escort is not amongst them; it is a saloon which could undoubtedly in its many and highly individual variants out-perform many of the cars we have included, but such high-priced Grand Touring exotica as the De Tomaso Pantera, Lamborghini Miura and Porsche Turbo demand the space; they are unequivocally 'sports cars' in the most modern sense.

From all this emerges one common theme. The pleasure of motoring, the joy of driving, is best served by those machines we call sports cars–and among their names we find some which by their special appeal, their competition success, their commercial success, or sometimes merely by their unusual technical interest or charm, must qualify for the 'classic' tag.

The arrangement of this book is basically chronological, with the cars in each year placed alphabetically, although there are one or two departures from this rule where it makes sense to group particular cars together, or where a particularly important car merits a two-page spread to itself. In what follows some enthusiasts may be outraged to find that their particular favourite has been omitted in favour of something they consider unworthy, but–like the definition of 'sports car', perhaps–it is all a matter of opinion . . .

The French Matra MS530 was typical of the high-performance sporting car of the seventies–roof and all.

'Gordon Bennett' Mercédès

1903, Germany

In the earliest days of motor-racing competition, the Gordon Bennett Cup was an annual International event contested by carefully chosen national teams of three cars each. The competition was French-dominated until 1902, when Selwyn Francis Edge's Napier was the sole runner to complete the distance from Paris to Salzburg. This British victory gave the Royal Automobile Club of Great Britain and Ireland the right to organize the 1903 event on their home territory. Since it was impossible to gain permission to close public roads on the mainland for motor-racing use, a large figure-of-eight course was chosen near Ballyshannon in what is now Eire.

The Germans produced an excellent racing car in their 90 hp Mercédès model, but just one month before the Irish race took place the Mercédès factory at Canstatt, near Stuttgart, was swept by fire. The racing department was gutted and its 90 hp cars were destroyed. A French Mercédès customer named Baron Pierre de Caters contacted Emile Jellinek, salesman-cum-promoter of the Daimler-built cars which were named after his daughter (Mercédès Jellinek), and suggested that stripped 60 hp production models such as his own might do the job in Ireland. The company took up de Caters's idea, stripped his car of unnecessary fittings and entered him in the Gordon Bennett.

Another 60 hp was called back from American sportsman Clarence Gray Dinsmore. His chauffeur Wilhelm Werner was, however, judged by the German club to be insufficiently qualified as a 'gentleman driver' to represent his country. The Belgian Camille Jenatzy therefore took his place.

At Ballyshannon the two stripped production Mercédès 60s ran very well, and Jenatzy scored Mercédès's first major International victory, averaging 49.2 mph over 327 miles.

Benz 50 hp

1908, Germany

The competitive events which first developed a true 'sports car. breed were organized in Germany before World War I. They were known as the International Touring Car Competition for the Herkomer Trophy. The first competition was sponsored by Professor Hubert von Herkomer RA, one of its prizes being a portrait of the winner painted free of charge by the Professor himself. It began in 1905 and was won by a hefty Mercédès 40 with more or less standard heavyweight coachwork. In 1906, however, coachwork regulations were relaxed and lightweight bodies appeared, to aid the cars' power-to-weight ratio and so improve their performance.

The Prince Henry Benz, notable for its flared wings — to become a trademark of later sports cars.

An 18/20 hp Horch of only 2.7-litres won.

Herkomer's last competition was run in 1907. One enthusiastic competitor in these pioneering events had been Prince Henry of Prussia – younger brother of the Kaiser – and he presented a trophy for a touring-car event to replace the Herkomer, open to four-seater vehicles, setting engine bore limits and excluding 'trade entrants'. The competition was known as the Prince Henry (actually *Prinz Heinrich*) trial.

The first Prince Henry Trial in 1908 saw exotic 'specials' appearing, far more exotic than Herkomer entrants had ever dreamed of; Horch, for example, ran stark four-seater cars with a cowl separating front and rear passengers, with flared wings but no doors. This type of body became known as the 'torpedo' type, the germ of the sports car. The event was won by Fritz Erle, a faithful Benz dealer and competition driver from Mannheim, driving a 50 hp Benz with rather less streamlined coachwork than the Horch competition but still with the flared wings and streamlined scuttle which were to typify later sports cars.

The Prince Henry Trial course included acceleration tests and against-the-clock hillclimbs, and Erle's big Benz was timed at 86 mph over a six-mile strip. In common with most other similar cars of the period it was a hefty four-cylinder with chain drive to the back axle, wooden-spoked artillery wheels and brakes only on the rear wheels and on the transmission shaft. It was built to go, not stop.

Stanley

1908, United States

This was among the earliest sporting cars to be built for sale in the United States of America. Steam cars had become widely popular for their silence and good pulling power, but apart from a few highly-tuned 'specials', they had all been fairly staid transport with hardly rib-crushing performance. More and more manufacturers turned away from steam towards the greater power instantly to hand with the internal combustion engine, but in America the Stanley brothers remained faithful to boiling water. For example, they had built an extraordinary World Land Speed Record contender named *Wogglebug* or *The Beetle*, which achieved the extraordinary speed of 127 mph on Daytona Beach in Florida before suffering a major accident which the driver, Fred Marriott, was fortunate to survive.

The Stanley company of Newton, Massachusetts, produced three roadster models, all of sporting character and each with stark but very light two-seater bodies. Their Model EX offered an 18-inch boiler and 10 hp engine, the Model H a 23-inch boiler and 20 hp engine and the Model K a 26-inch boiler and 30 hp engine. It was the median Model H which they delightfully named 'The Gentleman's Speedy Roadster'. Each of these models weighed in at under 1700 lb and their performance was truly amazing for 1906–08. The Model H in fact laid claim to the title of Fastest Stock Car in the World after being timed at 68.18 mph during a 15-mile handicap race on Ormond Beach, Daytona.

One of the Stanley twins, F. E., was once arrested for driving on the public road 'at nearly 60 mph' in one of the products which he and his twin F. O. drove so enthusiastically. The story goes that in Court he pleaded not guilty to the charge '. . . of going 60 mph. When I passed the officer my speedometer showed I was going 87 miles an hour!'. The judge evidently appreciated the case's promotional value and fined him all of five dollars.

Stanley

The brothers Stanley were loyal to steam longer than most of their competitors.

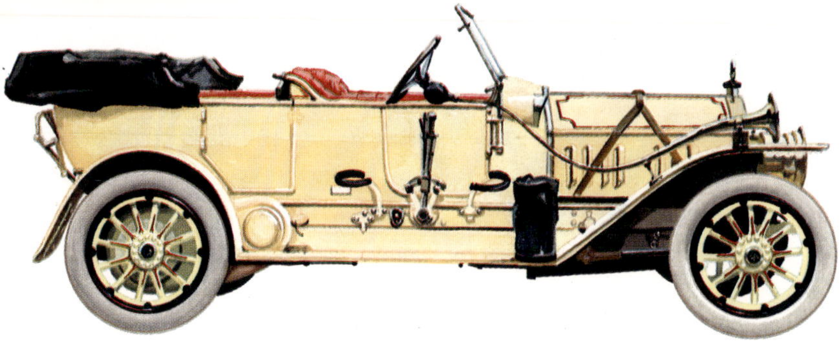

Isotta-Fraschini 100 hp

1911, Italy

Fabbrica Automobili Isotta-Fraschini of Milan, Italy, developed from a business partnership between Cesare Isotta and Vincenzo Fraschini. They began producing cars wholely to their own design in 1903 and were soon involved in competition in an attempt to prove their wares.

The 100 hp model was probably the most technically interesting high-performance production car of the period when introduced in 1910, and later modified in short-chassis form in 1913. It was a handsome beast with a tall and lengthy bonnet merging into a streamlined scuttle and high-sided, open-top, four-seat bodywork. A raked-back windscreen gave an impression of speed even when the car was stationary, and curvaceous wings merging into long running boards all looked very modern at the time.

Under the bonnet the Isotta-Fraschini was certainly ahead of contemporary practice. The 130 mm bore × 200 mm stroke engine gave a capacity of 10,618 cc and featured a single overhead camshaft operating four valves per cylinder with water pump and ignition magneto driven by skew-gears and shafts from the vertical camshaft drive. This engine delivered some 125 bhp. Whereas most of the large overhead-camshaft engines of this period had exposed valve gear and springs the Isotta had a neat cover to cowl in all such essentials. In fact with monobloc fixed-head cylinders and steel water-jackets, affixed by countersunk screws, and separate cylinder bases, it virtually forshadowed the huge $4\frac{1}{2}$-litre Bentley engine to come over a decade later. Isotta also boasted four-wheel braking at a time when front-wheel brakes were widely regarded as a dangerous excess. What was more, Isotta developed a rod control system for their front-wheel brakes which made them consistent and reliable in operation, allowing the power of their car's sophisticated engine to be employed to the full without fear.

Isotta built a little 27/80 sister version of their startling 100, using a 105 mm × 180 mm 6234 cc version of the big 10-litre engine. Both models built a remarkable reputation.

Bédélia Cyclecar

1913, France

Around 1910 there was a move towards inexpensive small-capacity cars with sporting performance, intended to provide a kind of minor competition league and 'fun-car' road use. The expensive big cars were after all available only to the extremely wealthy. Thus appeared the first of the cyclecar genre.

The cyclecars, baby four-wheelers, owed more to motor-cycle practice than to the motor-car world. In Britain the GN made its debut, while in France there appeared the Bedélia. Both were typical.

The Bedélia was developed by Bourbeau and Devaux, setting out with a V-twin engine though later offering singles to order. The

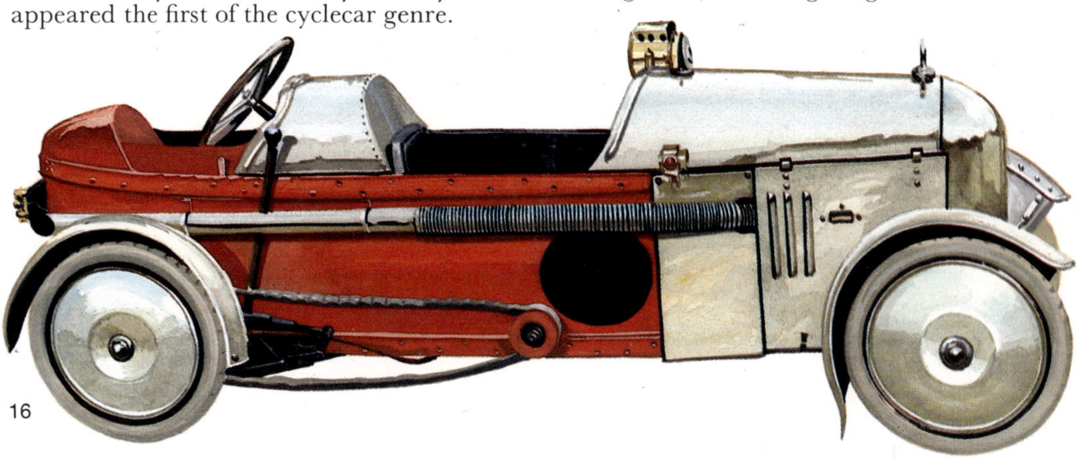

engine was connected by chain to a countershaft half-way along the frame, from which final drive to the back axle was by belts. But unlike the GN, which seated its occupants side by side, the Bedélia placed them in tandem in a slimline bodyshell. Steering was by wire-and-bobbin, swivelling the whole front axle in a dangerous manner which most major manufacturers had long since abandoned. But in the lightweight Bedélia the arrangement was good enough for many enthusiastic customers, and since these cyclecars were very cheap to buy, maintain and run, and could achieve around 45 mph as standard and anything up to a dizzy 60 mph when competition-tuned, many were sold. Bedélias did well in competition as well as on the open road.

Archie Frazer-Nash and H. R. Godfrey built a tiny 'special' powered by a 90-degree V-twin 1100 cc JAP engine mounted in line with their chassis, and soon attracted numerous customer orders for replicas. They developed their own engine and with a wooden frame made of ash, and chain-and-belt transmission, their resultant GNs weighed in at little more than 400 lb. Power-to-weight ratio was very favourable and they proved very fast for their size. Latter-day enthusiasts may think that a Morgan or Healey is fairly crude; but for hair-shirt motoring nothing could beat these early cyclecars.

Mercer Raceabout

1913, United States

The Mercer company was named after Mercer County, New Jersey, USA, where it built its cars. It was founded in 1909 as a successor to Roebling-Planche who had made early cars, and members of the Roebling family remained on the Mercer board throughout its peak years. Mercer's chief engineer Finlay Robertson Porter designed the Mercer Type 35 series, which was introduced in 1910 with the Raceabout as its most prestigious model.

The Mercer Raceabout used a modest four-cylinder T-head dual-ignition engine of only 4.9 litres delivering some 60 bhp. The engine was mounted within a very light and simple bodyshell with extremely sporting lines; a bolster tail tank was placed behind two confined bucket seats mounted on an exposed chassis platform with no sides to the cockpit and only an abbreviated engine cowl up ahead. With long flowing wings and running boards the 'Raceabout' tag really described the spirited look of the car. The only concession to weather protection was a circular 'monocle' screen which was available to bolt onto the steering column.

The Raceabouts were fine cars for their day, though their lightweight chassis frames proved susceptible to cracking as the miles mounted up, road ride was very hard and jittery, braking was indifferent at best and wind buffeting was a major problem. However, the Raceabout had all the flair and spirit which was to typify the sports car of a later age, and it sold very well indeed, eventually becoming one of the most sought-after pre-Great War antique cars.

Purebred racing versions of the Type 35 series brought Mercer to the forefront of American competition in these years and when in 1914 Eddie Pullen won the American Grand Prize race for them the whole works was given a half-day holiday. But in 1912 the family scion, Washington Roebling III, had been drowned in the *Titanic* catastrophe and Mercer missed his personal flair and drive. By 1918 the surviving Roeblings had died; and the Mercer company changed hands and its image faded.

'Brescia' Bugatti Type 13

1922, France

The 'Brescia' Bugatti of the twenties was a development of the sensational Type 13 pre-Great War Bugatti model which had performed so strongly in the 1911 'Grand Prix des Vieux Tacots' at Le Mans. The Type 22 became known as the 'Brescia' after its sweeping success in the International 'Grand Prix des Voiturettes' at Brescia in Italy in 1921; it had actually been tested experimentally as early as 1914 in slightly different form.

The modified Type 22 began to achieve fame as a production car in around 1922 and by 1926 it was effectively obsolete; but in those intervening years it was one of the fastest small cars on the market. Ettore Bugatti's overhead camshaft engine of 69 mm bore and 100 mm stroke, giving the four cylinders a total swept volume of 1496 cc, used his famous curved tappets of square section, plus a multi-plate clutch feeding a close-ratio four-speed gearbox. A touring set of gearbox ratios offered speeds of around 40 mph, 60 mph and 75 mph at 3800 rpm while in high-geared competition versions maximum speeds rose to 85 mph or more. This usually meant that peak engine revolutions could not be achieved in top gear before the engine ran out of steam, and the Brescia would run faster in third. The engines were quoted as giving some 40 bhp at 3800 rpm and in sporting trim they weighed around 1340 lb.

What was more, the Brescia Bugattis handled brilliantly for the period with positive steering and high cornering powers which attracted a fanatical band of enthusiast owners. Bugatti insisted that his cars should go and was never very interested in making them stop, so the Brescia's brakes were at best of the 'press hard and pray' variety.

In Britain Raymond Mays dominated sprint and hillclimb meetings in the early twenties with his two famous Brescias, Cordon Rouge and Cordon Bleu, both being progressively and highly modified to achieve 5800 rpm with reliability and proving capable of exceeding 90 mph with ease.

Morgan Aero

1924, Great Britain

H. F. S. Morgan founded his specialist motor company at Malvern, Worcestershire, in 1910 to produce a unique tricycle car. The single wheel was at the rear of a tubular chassis frame. Up front was a side-valve air-cooled V-twin motor-cycle engine of 1100 cc made by JAP. Transmission was by dog clutch and chains, steering was direct and there was independent suspension to each front wheel by sliding pillars and coil springs. There were two simple seats and with reasonable power and light weight this little three-wheeler was capable of considerable performance.

The Morgan proved to have road-holding superior to that of most machines in its class and it rapidly made its name and reputation in sprint and hillclimb events. In 1914 H. F. S.

Morgan introduced his Grand Prix model, the first competition Morgan to be catalogued, and soon afterwards a confined four-seater was unveiled, to become the 'family' model of the twenties.

By March 1919 the little factory at Malvern Link was turning out about 20 cars a week, and through the twenties various engines were used including MAG, JAP and Blackburne. Overhead-valve engines were introduced in 1923 and standardized by 1925. By 1927 the Morgan Aero (introduced with Blackburne engine in 1921) could achieve a genuine passenger-petrifying 80 mph while at more modest touring speeds would return 45 mpg. The Morgan was inexpensive to buy, maintain and run, and its high-speed performance and vast competition pedigree made it the object of almost fanatical love for many enthusiasts.

Lancia Lambda

1925, Italy

Vincenzo Lancia's Lambda model set new standards of handling in 1921–22. Lancia himself had been an electrifying racing driver to watch in the Heroic Age of racing, but after retiring to build production cars he steered clear of competition. He was nevertheless a constructor who enjoyed driving – and this showed in his products just as much as it does with enthusiast managements today, notably BMW and Opel in Germany.

The Lancia Lambda inherited a narrow-angle 'V' engine with overhead camshafts – though in four-cylinder form – from its immediate antecedent, the V8 Lancia Trikappa. The chassis-cum-body was constructed on unitary principles, way ahead of its time, and employed independent front suspension of vertical coil-spring type, as found on all Lancia models until 1956 and present within their range as late as 1963. The Lambda also featured such 'revolutionary' ideas as alloy cylinder blocks, pump cooling, full-pressure lubrication, vacuum fuel feed and four-wheel brakes – all rolled into the one package. Unitary construction with a propeller tunnel for the transmission shaft allowed the Lambda body to be built unusually low for its day, and with powerful cable brakes and 2120 cc engine giving 50 bhp at 3000 rpm the new Lambda was a remarkably exciting innovation. Its wheelbase was too long to make the most of its nimble handling potential but the model proved a great success.

The Lambda was built in eight series until 1931; for the seventh in 1927 engine dimensions were changed from 75 mm bore × 120 mm stroke to 80 mm × 120 mm (2.4 litres), and in 1928 the final models offered 92 mm bore 2.5-litre power units. The standard torpedo touring body was continued little changed through these years with a detachable hard-top available for conversion into a saloon. Early numbers could exceed 70 mph, the later ones 80 mph, but what was more important was their remarkable road manners: smooth riding, controllable and comfortable.

3-litre Bentley

1926, Great Britain

The 1919 London Motor Show was highlighted by the appearance of Walter Owen Bentley's prototype sports car, which carried his own name. W. O. Bentley had been British importer of the DFP car which pioneered aluminium pistons, and during World War I he had designed a string of BR rotary aero-engines. The first sports car carrying his own name had an 80 mm × 149 mm four-cylinder single-overhead-camshaft engine with fixed cylinder head and dual-magneto ignition; it delivered some 70 bhp in prototype form. In the 1922 RAC Tourist Trophy race on the Isle of Man the factory team of three flat-radiator Bentley 3-litres finished 2nd, 3rd and 5th, and in 1924 Duff/Clement won the prestigious Le Mans 24-Hour Grand Prix d'Endurance for the Cricklewood concern. In 1927 the Davis/Benjafield 3-litre Bentley, 'Old No 7', won again at Le Mans after being severely damaged in the extraordinary 'White House Crash' which wiped out the rest of the Bentley team. Up to 1929 Bentley 3-litre production totalled 1630 cars of which a large proportion survive as collector's pieces today.

In 1924 the model had received front-wheel brakes and Vanden Plas – the coachbuilding company – had also introduced their famous sports four-seater 'Speed Model' on this chassis. Red-enamelled radiator badges signified the 'Red Label' Speed Model short-chassis 3-litre Bentley; blue enamelling the early short and long chassis which could and occasionally did carry limousine coachwork. Green enamelling indicated the very special, short-chassis 100 mph type made in restricted numbers.

The classical 3-litre Bentley of the vintage years was noted for its high build and 'Gothic' radiator.

Of all the vintage Bentleys the 3-litre is perhaps the most graceful and least truck-like with its tall radiator profile and spare

VSCC meeting at Silverstone 3-litre Bentley on modern, small-diameter wheels.

and subtle lines. Its simple 'four' carried 16 valves in pent-roof combustion chambers demonstrating their designer's faith in the pre-World War I Peugeot Grand Prix cars, as did the stroke-to-bore ratio of some 2:1. The Bentley's single overhead camshaft and valve gear was similar to that which W.O. had studied on the 1914 Grand Prix-winning Mercedes, whose engine was in many ways decades ahead of its time. Bentley recognized its merit and improved upon it.

He designed reliability into what Ettore Bugatti referred to caustically as 'the world's fastest lorries'. He spent hours refining his drawings and production methods to ensure that features such as crankshaft support, cooling and overall ease of maintenance were honed to the ultimate degree. On such care and attention to detail he forged Bentley's matchless sustained performance on road and track.

While the Speed Model could cruise at 85 mph in standard form and be tuned to produce 100 mph virtually all day long, it could equally well pull at 8 mph in top gear, and average a fuel consumption of 21 mpg. No wonder the 'old 3-litre' proved such a magnificent beast during its heyday.

At Le Mans for the 1926 24-Hours race, Bentley fought hard to prevent Lorraine winning for the second year in succession. The British marque had won in 1924 but this 3-litre model crashed when third with only 20 minutes to run. It was shared by 'Sammy' Davis and Dr. Benjafield; in 1927 'Old No 7' won the French classic after surviving the infamous 'White House Crash', again shared by this experienced pair.

Vauxhall 30/98 OE

1926, Great Britain

Before World War I the Vauxhall Motor Company was very active in competition and gained a great reputation under the technical guidance of its youthful and dynamic Chief Engineer, Laurence Pomeroy. With the backing of chief executive Percy Kidner, Pomeroy had developed the Prince Henry three-litre and later four-litre high-performance touring model in 1912–13. The larger engine did not give Vauxhall the competition edge they envisaged, and so the 25 hp four-litre engine was given a bore and stroke of 98 mm × 150 mm to provide 4525 cc: the resultant car was named the 30/98.

John Higginson, a well-known textile engineer from Stockport, Cheshire, scored many early sprint and hillclimb successes for the new model and about a dozen replica 30/98s were delivered before the Great War demanded all production capacity.

In the early twenties the Vauxhall 30/98 was indisputably England's premier sporting car. Developed from the stark pre-war competition specials, it carried full road equipment and eventually the famous Velox four-seater body—this combined four seats with slim-line looks of unusual grace and subtlety. The engine was still the Edwardian fixed-head unit with side valves and a single Zenith carburettor, but it delivered a reliable and honest 90 bhp at 2800 rpm and Vauxhall could claim to have the fastest car then available in standard production. The stock model could achieve 85 mph on the public road, while with a racing body a 100 mph Brooklands lap was guaranteed. Several owners

Vintage sporting profile: the Vauxhall 30/98 with four-seater touring body.

Classic Vauxhalls of the vintage era, recognizable by their subtly fluted bonnets, built upon the design and engineering expertise of Laurence Pomeroy and drew upon a background of pre-Great War competition experience. They were aimed not only at the British RAC Tourist Trophy but also the mighty Grand Prix de l'ACF.

insisted on putting this claim to the test, and were satisfied with the result. The car was high-geared, as were most of its genre, and 65 mph cruising was very comfortable and relatively quiet at a lazy 2000 rpm.

The Vauxhall catalogue described the model as 'a very refined fast touring car capable of high average speeds and suitable for competition work'. The manufacturers were justified in claiming refinement for the car; the rest of this description still holds good as the standard by which sports cars of the world should be judged.

In order to achieve greater flexibility throughout the rev range the E-Type engine was modified in 1922 as the 'OE 30/98', offering a shorter stroke and pushrod-operated overhead valves. For 1923 it became a high-speed engine, revving to 3500 rpm, at which speed it produced 112 bhp, compared to the E-type's 90 bhp at 2800 rpm. A counter-balanced crankshaft improved matters still further in 1926 when the boat-tailed 2–3-seat Wensum body style was introduced. It was handsome and rakish, but never replaced the Velox four-seater in the public's esteem.

By 1926 Vauxhall 30/98s had captured something like 13 firsts, 19 seconds and seven thirds at Brooklands Motor Course alone, and OEs added 14 firsts, nine seconds and seven thirds there. Vauxhall kept in close touch with customers who raced their products until in 1926 the American General Motors Corporation bought the English company; the marvellous 30/98 remained in production for just as long as it took to convert the plant for the production of more mundane vehicles of greater mass appeal.

Vauxhall 30/98 at a British Vintage Sports Car Club driving-tests meeting in recent years; a big car to throw around the pylons but more nimble and controllable than it looks.

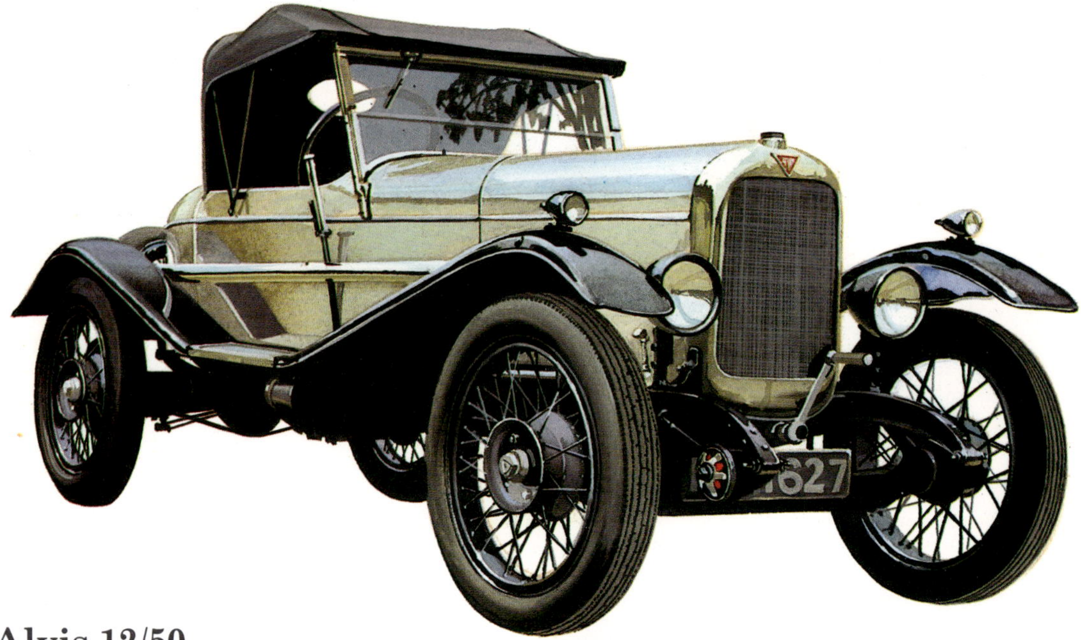

Alvis 12/50

1927, Great Britain

T. G. John of Coventry, England, founded his Alvis car-manufacturing company in 1919. In 1923 John's chief engineer Captain G. T. Smith-Clarke produced the pushrod overhead-valve 12/50 engine which was quieter and more powerful than its immediate predecessors and coincidentally added remarkable strength and reliability. Alvis won the Brooklands 200-Mile race of 1923 with this engine in tuned competition form, and touring 12/50s entered production with 1598cc engines. The Super Sport versions had a shorter stroke giving only 1496cc, but allowing higher revolutions and more power. While the touring cars could run up to 60–65mph, the tuned Super Sports could achieve well over 70mph on the straight and level. Works driver Major C. M. Harvey's victory in that 1923 200-Miles race had formed the perfect base from which Alvis could launch their new model, for he had averaged no less than 93mph for the distance, and his special car's fuel consumption was a remarkable 24mpg.

While the company endured various vicissitudes during the mid-twenties, the 12/50 Alvis cars went on to establish a remarkable popularity and they continued in production until 1932. During that period various differing body styles were adopted, the chassis was stiffened, the separate engine subframe was replaced and, in 1924, front-wheel brakes were adopted.

1925 Alvis 12/50 'Duck's Back' in the Land's End Trial.

Although the 12/50 was fairly expensive during its run it sold in larger numbers than any other comparable British 1½-litre car. Many survive and are in regular use today. They are a practical proposition for the impecunious 'Vintagent' and still prove themselves to have been well made to a sound and reliable original design.

'Brooklands' Riley Nine

1927, Great Britain

The Riley Cycle Company of Coventry, Warwickshire, built their first car as a small single-cylinder voiturette with belt-drive. It did not enter production. In 1900 they began production in a modest way, building motor tricycles; tricars with simple engines continued in production until 1907. Later models used a 1034cc V-twin engine and it was this unit which powered the first four-wheeled Riley cars. Through the twenties Rileys grew in sophistication and performance and in 1927 there was even a supercharged version of the Redwing available.

Then Percy Riley introduced his advanced and sophisticated Nine model. The prototype Riley Nine was announced in the summer of 1926 and justifiably caused a sensation. Its engine was a 1087cc, 32 bhp four-cylinder unit in which the camshafts controlling valve opening were situated high in the engine flanks but not overhead, actuating the inclined valves through short pushrods, combined with a cylinder head featuring hemispherical combustion chambers–known as the 'PR Head' after Percy Riley himself. This unit was to remain the basis of Riley production until as late as 1957. By the end of 1926 the great Welsh driver-engineer J. G. Parry Thomas was working on the Brooklands sports version of the Riley Nine, and after Parry Thomas died attacking the Land Speed Record at Pendine Sands in March 1927 the work was continued by Reid Railton.

The Brooklands Nine became the epitome of the British sports car, with low-slung and rakish lines, an almost horizontal steering column and flaring front wings. The specialist firm of Thomson & Taylor assembled the first few cars actually at Brooklands, but Riley later took over production themselves. With high-compression pistons and high-lift cam profiles the Brooklands' power was boosted from the Touring Nine's 32 bhp to 50 bhp at 5000 rpm. The Brooklands Nine could achieve 80 mph easily, making it a true match for the French Salmson and Amilcar opposition, and it was to achieve a long and noble record of competition success.

Non-standard approach to the Ards TT pits in Ulster, demonstrated during the 1928 race by this Riley Nine.

Mercedes-Benz 38/250 'SS'

1928, Germany

The great German marques Mercédès and Benz merged in 1926 and in that year former Austro-Daimler designer Ferdinand Porsche unveiled for them their Model K. This was an overhead-camshaft 6.2-litre six-cylinder sports car in which full depression of the throttle pedal engaged a supercharger to force-feed the engine. Performance and sound volume increased jointly when the 'blower' cut in, 0–50 mph acceleration in around 30 seconds being improved to 20 seconds by flooring the throttle in this way. The car was good for 95 mph and was a prime example of sheer engine size and then supercharging being used as a means of improving performance.

The high-rigged K models proved difficult to handle, and in 1927 the S or 'Sports' variant appeared, with a lowered chassis line and a 98 mm × 150 mm 6789 cc engine; it offered 120 bhp at 3000 rpm unblown, or 180 bhp with the supercharger engaged. The K's brakes had been improved for the 105 mph 'S' by the use of servo-assistance, and S models were placed 1–2–3 in the 1927 German Grand Prix at Nürburgring.

In 1928 the 'SS' appeared, using a 100 mm × 150 mm 7069 cc twin-carburettor supercharged engine delivering 200 bhp; in stock form under favourable conditions this $2\frac{1}{2}$-ton monster could achieve a genuine 110 mph.

However, the massive chassis of the SS was intended also to accept staid touring coachwork and the 'SSK' – K for *kurz* (short) – was developed as a lightweight short-chassis competition variant with strictly two-seat bodywork and a larger-capacity supercharger. In one of these mighty cars Caracciola won the 1929 Tourist Trophy race at Ards. For 1930 a handful of lightened 'SSKL' – *leicht* (lightweight) – models were produced delivering around 300 bhp and capable of as much as 130 mph.

The 'great white Mercedes' achieved major and minor competition success all over Germany and Europe into 1932 and despite their brute force, ignorance was never resorted to in other aspects of their design. They were truly great, classic sports cars, and the Caracciola/Werner performances in the Mille Miglia around Italy and Le Mans 24-Hour races bear testimony to this fact.

Rudolf Caracciola breaking the Semmering mountain-climb record in 1928 in Austria, driving his works Mercedes-Benz 'SSK'.

Francis, the Earl Howe, storms Shelsley Walsh hill-climb in his Mercedes-Benz 38/250 'SS' – quite a sight for the large crowds which flocked to the English venue in the early thirties.

All the K- and S-series cars used the same supercharged single-overhead-camshaft six-cylinder engine varying only in minor details and in capacity. Block and crankcase were a single light-alloy casting, which was quite daring in those days. The camshaft and all the auxiliaries were shaft-driven from the crank. Despite its size and apparent complication this was a fairly simple engine upon which to work, and has been described as a stolid workhorse rather than a highly-strung thoroughbred.

The cars' greatest fault was their inability to stop – their brakes were inadequate relative to their great power, weight and straight-line performance. Nevertheless, even the long-chassis cars handled and cornered well for their time. Much of these cars' development may be credited to Dr Hans Nibel, Benz chief engineer prior to their 1926 merger with Mercedes, and his refinement of Dr Porsche's original design certainly made the series the immortal success they became. Nibel went on to produce other classic Mercedes, the production 500K/540K/770K series with coil-sprung rear swing axle suspension, the rear-engined Type 170, and the original W25 Grand Prix Mercedes of 1934 – the dawn of Mercedes-Benz's greatest era.

▼ *Mercedes-Benz 'SSK', devised by Dr Porsche of later Auto Union and wartime tank fame, and the apotheosis of 'vintage' sports-racing-car ferocity, looked every inch the monster it could be to drive.*

Tulipwood Hispano-Suiza

1929, France

The Swiss engineer Marc Birkigt was responsible for the high-quality Hispano-Suizas which were among the exotic supercars of their day, equivalent to the Lamborghinis and Ferraris seen now.

The first Hispano-Suiza was built in 1904 in Barcelona, Spain, and was actually a Birkigt-designed 20 hp Castro car renamed. A range of beautifully built and very sophisticated Hispanos flowed thereafter from Birkigt's drawing board. All were conventional but highly developed and very expensive cars and by 1907 they were found in the garages of King Alfonso XIII of Spain. In 1910 an Hispano-Suiza voiturette won the important Coupe de l'Auto voiturette competition in France, and from 1911 Hispano production began in a new French factory in Levallois-Perret. This very modern plant produced, in 1919, the stunning H6b to cater for France's luxury car market, very much larger than that available in Spain.

The H6b was designed from the ground up as the ultimate road transport for the wealthy, using an aero-engine-derived light-alloy six-cylinder power unit of 6.6-litres with single overhead camshaft, two valves per cylinder and a seven-main-bearing pressure-fed crankshaft. With 135 bhp at 3000 rpm in a light yet rigid chassis, and using four-wheel brakes designed to control an 85 mph top speed with ease, the H6b qualified as one of the earliest Grand Touring cars – its performance put it into the sporting league despite its luxurious furnishing and equipment. It did not have the quiet refinement of a Rolls-Royce, yet it could out-perform virtually any Rolls with one wheel tied behind its luggage trunk, and Hispano-Suiza effectively pursued a one-model policy based on it for many years.

Salmson 10-20 hp

1929, France

The Emile Petit-designed cars from the Société des Moteurs Salmson of Billancourt, France, made a tremendous reputation for themselves in the 1920s. Salmson had been renowned for their aero engines during World War I and in 1919 turned to car manufacture by taking a licence to produce British GN cyclecars. They built 3000 GNs; then, in 1921, their own first models appeared with a flimsy cyclecar-type chassis and an 1100 cc monobloc four-cylinder engine, in which a single valve-gear control rod per cylinder operated as a pushrod to open the exhaust valve and reversed duty as a 'pull'-rod to operate the inlets.

For the Cyclecar Grand Prix of 1921 Emile Petit introduced a twin-overhead-camshaft version of this engine and Lombard won the race with it, and was then second behind a GN in the 200-Miles Race at Brooklands. In 1922 Salmson cars were placed first and second in both these events; in 1923 they won their class at Le Mans and went on to achieve enormous competition success.

From 1925 the twin-cam 1100 cc Salmson sports cars took a grip on their competition class and with a cowled radiator from 1926 their classical shape became synonymous with sporting achievement. All the sports models were available with balloon tyres, front-wheel brakes and back axles with no differential. The Grand Prix model had a three-speed gearbox and splash lubrication for its two-bearing crank to achieve 70 mph top speeds, while the GP Special used Cozette supercharging, four speeds and pressure lubrication, and was capable of 90 mph or more.

Like Amilcar, however, Salmson suffered when MG's low-priced Midget became generally available and under-cut them in the French small-sports-car market. Salmson headed up-market, as did Amilcar. After surviving World War II, Salmson lingered on until 1957, when it was acquired by Renault.

Austin Seven Ulster

1930, Great Britain

Austin

Of all the touring cars which spawned successful competition variants perhaps the diminutive Austin Seven was the most humble.

With its FN motor-cycle-derived engine producing only 13 hp and a chassis frame similar to that of the Peugeot 'quad', this unprepossessing little model at least offered four-wheel brakes. At only 747 cc the Austin Seven was eligible for the 750 cc competition class, in which it faced little opposition. Herbert Austin saw the opportunity for competition prestige on which to sell his production model, and as early as 1924 his company offered a Sports Model Seven with pointed tail, 960 lb weight and 52 mph.

E. C. Gordon England 'breathed' upon the Sports Model Seven to great effect, producing the 55 mph 'Cup' model, and then a very spartan but rapid road-racer which he named the Brooklands. A Cozette-supercharged Super Sports model was introduced by Austin themselves in July 1928, offering 33 bhp and no less than 70 mph. In the 1929 Tourist Trophy at Ards in Ulster Captain Archie Frazer-Nash finished third on handicap in one of these cars behind Rudolf Caracciola's 7.1-litre Mercedes-Benz and Giuseppe Campari's 1.5-litre Alfa.

In 1929 the Austin Seven Ulster was made available to the public in supercharged and unsupercharged forms, and in the words of one sports car expert, 'serious attention was given to the normally unconventional roadholding of the car and the brakes were also caused to work' – coupling together the four-wheel brakes, which had formerly been activated up front by the foot pedal and at the rear by a hand-brake. The little 747 cc engine would rev to 5000 rpm and with a close-ratio three-speed gearbox and weight as low as 1200 lb top speed was way above 70 mph.

Amherst Villiers' supercharger on the nose typified the 1930–31 'Blower' Bentley.

$4\frac{1}{2}$-litre 'Blower' Bentley

1930, Great Britain

In 1927 the Bentley 3-litre was developed into a $4\frac{1}{2}$-litre variant – still with four cylinders – and development pushed its output up ultimately from 100 bhp to 130 bhp. In 1928 Barnato and Rubin drove a 'four-and-a-half' to victory at Le Mans to underline Bentley's grip on this trial of endurance and speed.

Meanwhile Sir Henry 'Tim' Birkin decided that to achieve optimum performance from W.O.'s design, supercharging should be adopted. The original designer never approved of supercharging as a method of obtaining increased power, always following the dictum that 'there is no substitute for cubic inches' – when he required more urge he preferred to enlarge the engine capacity.

The Hon. Dorothy Paget financed Birkin's plan for a supercharged $4\frac{1}{2}$-litre Bentley and the cars were modified in her premises at Welwyn. The resultant 'Blower Bentley' developed a staggering 240 bhp at 4200 rpm but proved unreliable and incapable of translating that power into race victories. The supercharger operated well at consistently high speeds, such as those achieved on the Brooklands Outer Circuit banking, and there Birkin's very special single-seater version excelled.

A total of 54 'Blower' Bentleys were built, the vast majority for sale as road cars.

We cannot leave these great sporting Bentleys without mentioning two other models of these late 1920s. In 1926 W. O. Bentley had aimed his sights at the luxury market by introducing a $6\frac{1}{2}$-litre six-cylinder chassis upon which the great coachbuilders of the day could weave their magic. Unfortunately the sporting image of the 3-litre Bentley was looked at rather disdainfully by the luxury coach customer, and the $6\frac{1}{2}$ was more successful in 1929 in the guise of the 180 hp 'Speed 6' – for many enthusiasts the best Bentley ever made, bringing the company its fourth and fifth Le Mans victories, in 1929 (driven by Barnato/Clement) and 1930 (Barnato/Kidston).

The depression year of 1930 was to bring the collapse of Bentley Motors, and their largest model – the 8-litre six-cylinder car – was cut short in its prime, only 100 being sold before take-over by Rolls-Royce and the loss of Bentley's independent identity.

OM

1930, Italy

SA Officine Meccaniche of Brescia, Italy, was a locomotive-constructor which absorbed the Züst concern in 1918 and built the Züst 25/35 hp car into 1923. OM introduced their own parallel designs, by Barratouché.

In 1920 a four-cylinder side-valve tourer was joined by the Tipo 665 Superba, a 1991 cc six-cylinder. While the little 'four' had no sporting claims, the two-litre built a surprising motor-racing reputation despite side valves and, normally, a single carburettor.

In 1926 at Le Mans, six-cylinder OMs came fourth and fifth, and in the first Mille Miglia race, in 1927, OM cars came first, second and third. In the 1928 Mille Miglia an OM was second behind an Alfa Romeo 1500. In 1928 Barratouché introduced a low-chassis OM which at least looked sporting and in 1929 its standard six-cylinder engine was enlarged to 2.2 litres. With a Roots supercharger this unit delivered 60 bhp at 4000 rpm. For 1931 its stroke was lengthened to provide a displacement of 2.3 litres, but no OM could now match the contemporary Alfas and Bugattis.

Talbot 105

1931, France

During the 1930s the London factory of Talbot Motors was directed with unerring technical sophistication and skill by Georges Roesch. He had saved the company in 1926 by his introduction of the touring 14/45 Light Six model, which suffered only from being grossly overweight for its 1660 cc engine. In 1930 an enlarged 2.2-litre version was introduced, which was known as the Talbot 90, and this became renowned for its smoothness and silence, its exquisitely made and carefully inspected seven-main-bearing crankshaft, and its excellent road manners up to a top speed of at least 80 mph.

Fox & Nicholl Ltd, the London Talbot agents, fielded a beautifully prepared team of lightened sports-racing '90s' from 1930 to 1933 and with 93 bhp at 4500 rpm in continued smoothness and silence they finished third and fourth behind the Bentleys at Le Mans in 1930, and won their class in the TT, Irish Grand Prix and Brooklands 500-Miles Race. Talbot found themselves once again in the front rank of high-quality sports cars and in 1930 Roesch introduced the 2969 cc 75 mm bore × 112 mm stroke Talbot 105 model giving 100 bhp as standard or 140 bhp in competition tune. Performance was outstanding for an unsupercharged car, particularly one of such refinement.

In 1932 the team of Talbot 105s carried off the team prize in the Alpine Trial, and one car was third at Le Mans in 1932. Mike Couper's special Brooklands version (later fitted with a 3.3-litre '110' engine) achieved over 135 mph and lapped Brooklands at 129.7 mph.

When Sunbeam collapsed in 1935, Talbot – a related company – was dragged to its knees, and the Rootes Brothers takeover resulted in the sweeping aside of all the standards which Roesch had set for himself and his marque, and of those to which his faithful work-force had built the cars.

1931 Talbot 105, designed by the famous Georges Roesch.

Bugatti Type 55
1932, France

Bugatti progressed with the Brescia Type 22 model from strength to strength during the twenties, their classical cars excelling in everything from Grand Prix competition through endurance sports-car and touring-car racing to outright production road use. In the Type 55 model Ettore Bugatti produced one of the finest sports cars in history. This model was powered by a variant of the Type 51, 2.3-litre engine with supercharging and twin overhead camshafts; this was a Grand Prix racing engine, installed slightly de-tuned in a Type 54 Grand Prix racing chassis. As a racing project the Formule Libre Type 54 had been shelved after proving itself good for 140mph or more; but it suffered from poor weight distribution and the deleterious effects which that has on handling and roadholding behaviour. The 4.9-litre engines earmarked for the Type 54s had been diverted to the Type 50T touring models, whereupon the Type 55 came into being to absorb the chassis design, with the smaller and lighter 'two-three' straight-eight-cylinder engine installed.

Topped by beautifully profiled flaring-wing bodywork, the Type 55 Bugatti made a glorious roadster with all the scintillating performance and road manners one might expect from such direct Grand Prix parentage. In first gear the model could top 50mph, in second it was good for 75mph, it could clip 100mph in third and achieved as much as 115mph in top gear–in absolutely standard tune. It accelerated from 0–60mph in around 11 seconds, which is sufficiently good to see off many 'sporty' cars in the seventies!

The Type 55 has been described as a true connoisseur's Bugatti with the reservations that it was noisy–as one might expect, and that it was also very stiffly suspended and therefore gave quite a hard ride. With the Type 57 series which followed, *le Patron* set out to put these 'shortcomings' right.

Le Mans Stutz
1932, United States

It was in 1914 that Harry C. Stutz's Indianapolis-based company introduced what became possibly the most famous American motor car of the vintage period–the Stutz Bearcat.

The Stutz sporting car offered the customer a choice of six- or four-cylinder engines, both very large; the former displaced 377 cubic inches (6178cc) and the latter 390 cubic inches (6391cc). With a top speed up around 85mph and rakish styling with, it was said, 'everything that wouldn't rust, including the passengers, out in the open', the Bearcat creamed off much of the former Mercer Raceabout market. In 1915 the works racing Bearcats were fitted with overhead camshafts and four valves per cylinder, even though Harry C. Stutz rapidly lost interest in his competition programme and in fact sold his company in 1919.

From 1920 the Stutz marque continued with cars based on the pre-war designs, improving them in detail from year to year. In 1924 the Speedway Six was introduced as the first overhead-valve pushrod-engined model, and in 1926 the first overhead-

camshaft, eight-cylinder emerged as the Series AA. Stutz resumed racing in 1927 and the AA was extensively modified to suit. The Stutz Black Hawk (named after the Indianapolis factory) dominated American stock-car racing in this period while Miller track-racing cars were entered under Stutz's name and had a fine season in 1928. That year Brisson's Le Mans Stutz was a Model BB, bored out and modified from the AA, and in 1929 the BB was replaced by the further bored-out Model M. A team of three returned to Le Mans in 1929, while Brisson's additional car had a Roots-type supercharger mounted between its front dumb-irons. Two of the Stutzes split their fuel tanks, a third placing fifth behind the Bentleys.

For 1930 Chief Engineer Charles R. Greuter produced a cylinder head with twin overhead camshafts and four valves per cylinder; two Stutz DV32s thus-developed ran at Le Mans, both retiring. Still the public thought of any sporting Stutz as a 'Bearcat' and in the Depression new Bearcat and Super Bearcat models were launched which regrettably failed to save this classic marque.

Alfa Romeo Tipo 8C-2300

1933, Italy

Of all vintage and post-vintage sports-car manufacturers, perhaps Alfa Romeo of Milan have the finest reputation. Their products are the most sought-after and most valuable today. ALFA was established in 1910 as the Anonima Lombarda Fabbrica Automobili by Cavaliere Ugo Stella, who had formerly controlled the Italian branch of the French Darracq concern. The early ALFA cars were designed by Ing. Giuseppe Merosi, and in 1915 the company was taken over by an industrialist named Nicola Romeo. After the Armistice the cars were marketed as Alfa Romeos. Despite incredibly modest production levels Alfa Romeo were very interested in competition, starting in 1914 with an unsuccessful Grand Prix car. This interest flourished during the twenties; in 1924–25 the Alfa Romeo P2 Grand Prix cars dominated the sport and Alfa became World Champion Constructor.

Vittorio Jano, ex-Fiat, became Technical Head in place of Merosi in 1926 and his first touring and sports car designs gained instant respect–employing single- and twin-overhead-camshaft six-cylinder engines in 1500 and 1750cc forms. When supercharged these cars dominated the major sports-racing events of 1928–30 with the exception of Le Mans, and 'the seventeen-fifty' Alfa became recognized as a great classic design. From 1931 to 1934 victory at Le Mans fell to Jano's enlargement of the 1750 theme–this model being the immortal 8C-2300 with straight-eight engine arranged in two in-line blocks of four cylinders divided by a central drive for the twin overhead camshafts.

With 2336cc capacity, a minimum of 130bhp at 4900rpm and a racing norm of 178bhp, the 8C-2300s were available in chassis lengths of 10ft 2in (the Le Mans model) or 9ft (Mille Miglia), and engines were occasionally taken out to 2.4 or even 2.6 litres. While a top speed of 105–110mph was standard, the very best Mille Miglia two-seater could achieve 125mph, a tremendous velocity for the time. In 1933 8C-2300s were placed 1-2-3 at Le Mans during their four-year domination of that prestigious event. But the model was expensive to produce, and only 188 or so were made–many of which happily survive today.

MG K3 Magnette
1933, Great Britain

Putting his back into the job – Eddie Hall urges on his MG K3 Magnette on the Isle of Man in the early thirties.

MG is possibly the most famous of all sports-car marques as far as the general public is concerned, the initials being taken from Morris Garages, the company in Oxford from which William Morris (later Lord Nuffield) developed his industrial empire. The Garages were managed by Cecil Kimber and when, in 1920, William Morris's manufacturing company discontinued production of its own sporting version of the Morris Cowley, it was the Garages who began experimenting with different bodies and engine-tuning on Morris rolling chassis.

In 1924 Kimber produced a version of the 1.8-litre Morris Oxford which he fitted with a mildly tuned engine, an attractive aluminium body and up-rated suspension. By 1927 this MG was called the 14/40 hp in honour of its improved engine output compared to the standard Morris 14/28. In 1928 Morris introduced the $2\frac{1}{2}$-litre six-cylinder overhead-camshaft 'Six' and Kimber developed this with redesigned block and cylinder head to power his own MG 18/80 hp. This model was further modified in 1930 with the fitting of a four-speed gearbox and stiffened chassis frame to form the 18/80 Mark II.

Meanwhile the Morris Minor-based 847 cc MG Midget had made its mark. This M-Type MG was raced widely from 1930. The Abingdon-based company became synonymous with sports-car competition, and its products in both works team and private hands were hugely successful.

Kimber turned his attention to the 1930 Wolseley Hornet small six-cylinder and developed it into the MG F-Type Magna. This was offered in sports and touring versions with the 1271 cc engine; from the Magna there was developed in 1932 the K1 Magnette saloon, followed rapidly by the K2 two-seater, and then the supercharged competition K3 Magnette 'racer'.

This remarkable 1086 cc 140 bhp sports-racing car won its class in the 1933 Mille Miglia and was driven to overall victory in the year's RAC Tourist Trophy at Ards by the Italian ace Tazio Nuvolari. K3s became universally respected as potent competition cars and excellent high-speed roadsters. Many were converted into true single-seater racing cars and the originals which survive today are highly prized and very valuable indeed.

Aston Martin $1\frac{1}{2}$-litre Mk II

1935, Great Britain

Lionel Martin and Robert Bamford built their prototype car in 1914, deriving its name from Martin's and that of the Aston Clinton hillclimb venue. It used a 1.4-litre side-valve Coventry-Simplex engine mounted in an Isotta-Fraschini voiturette chassis; production commenced in 1922 and 60 or so of these cars and their immediate derivatives were built until 1925, many achieving considerable competition success.

The company was re-formed after 1925 and an overhead-camshaft $1\frac{1}{2}$-litre model was designed by A. C. Bertelli, was tested in an Enfield-Allday racing-chassis frame, and entered production at Feltham, Middlesex, in 1927. The car's bodies were designed by Bertelli's brother Enrico and by 1929 the models were well-established. In 1931 a brief marketing link was forged with Frazer Nash and in 1933 the company came under the financial control and direction of R. G. Sutherland.

Bertelli and his partner W. S. Renwick had introduced their 'International' model in 1930, so called because it was equipped for road racing under International rules 'anywhere in the world'. The car employed dry-sump lubrication as standard and a close-coupled two/four seat body. 'Internationals' were made on chassis of two wheelbase-lengths, one 8ft 6in and the other 10ft, and sports, saloon, tourer and coupé bodies were all adopted. In 1931 Astons had come second in their class in the Brooklands Double-12 Hour race and in 1931–32 won the Biennial Cup at Le Mans. In 1933 a Le Mans Replica model was catalogued using unit construction of engine and gearbox for the first time.

In 1934 a long chassis appeared carrying full four-seat tourer, coupé and saloon bodies, and in 1935 the Mark II emerged, identified by its chromed radiator grille and featuring a modified engine, still of 1495cc but with improved crankshaft and valve-gear design, and giving 73bhp. When the marque won the TT team prize in 1934 the stage was set for a super sports model named the Ulster in 1935. Offering 80bhp at 5250rpm and with a top speed of 102mph, this was a true sports car which did much to confirm Aston Martin's ever-growing reputation.

1936 $1\frac{1}{2}$-litre Aston Martin Mk II low-chassis model in the JCC Brooklands Meeting of March 25, 1939.

Fiat Ballila

1935, Italy

The colossal Fiat company derives its name from the initials of its full title, Fabbrica Italiana Automobili Torino. It was founded in Turin in 1899 by Giovanni Agnelli, di Bricherasio and the Count Carlo Biscaretti di Ruffia. Fiat went racing widely until the mid-twenties while building a wide range of production vehicles of all types. In 1932 the advanced small Fiat was introduced as the Tipo 508 Balilla, with a short-stroke, four-cylinder engine. This three-main-bearing unit had a bore and stroke of 65 mm and 75 mm respectively – 995 cc. Essentially the Balilla, which had hydraulic brakes, was a modest little touring car, but sports versions accounted for about 1000 of the total 113,000 produced and they gave a good account of themselves in up-to-1100 cc-class sports-car racing.

The sports Balilla engine used pushrod overhead valves and delivered about 36 bhp; it was installed in a sketchily bodied vehicle whose weight was little more than 1340 lb. Performance was actually on a par with the far more august Lancia Aprilia, and the little Tipo 508S was virtually the only small Continental sports car in opposition to the British-built babies. It sold quite well and raced in many countries of Europe. The 1933 Mille Miglia saw special Balillas running with prototype overhead-valve heads and four-speed gearboxes made by the Siata tuning firm, before the overhead-valve engines came as standard to the production models in 1934.

Two types of Balilla sports car were built, sharing the same engine. One was the Spider Normale with flowing wings and the other the Spider Corso with cycle mudguards. Fiat also produced a Berlinetta Aerodinamica, a futuristic little fastback coupé. English-marketed Balillas had home-grown bodies with pronounced tail fins and the cars did very well indeed in rallies. In France Amédée Gordini's tuning company made its name with the French-built version known as the Simca.

Italy's much-loved little Fiat Ballila won fans throughout the world for its pert good looks and surprising performance.

Frazer Nash

1935, Great Britain

When Captain Archie Frazer-Nash left GN in around 1924 he tried building conventional family touring cars with shaft drive, but his sporting instincts triumphed and the Frazer Nash car (without the hyphen, notice) reverted to type. His prototype used GN-like dog-clutch gearchange with separate chains driving each of the three forward speeds, a solid rear axle, high-ratio steering and virtually solid leaf-spring suspension. The original Plus-Power engine was replaced by a side-valve water-cooled four-cylinder Anzani of $1\frac{1}{2}$-litres capacity, giving about 40 bhp. The Frazer Nash in this form attracted widespread popularity.

Such changes as there were centred mainly upon the power units adopted; an overhead-valve Meadows engine with 50 bhp and overhead valves replaced the Anzani from 1929, and was allied to four-speed transmission. The 'Nash was fast and nimble and could out-corner virtually anything on the open road, retaining this competitive edge far into the thirties. The Meadows-engined four-speed 'Boulogne' model of the early thirties could lap Brooklands quicker in third gear than in top – at around 80 mph. From 1934 the com-

pany's own engine (designed by Albert Gough and known by his name) was adopted; this unit displaced 1496cc and used a single overhead camshaft to produce 60bhp. Only 26 Gough-engined models left the Isleworth factory, between 1934 and 1938, and its bugs were never properly ironed out. In that same period a smoother six-cylinder Blackburne engine of 1657cc with twin overhead camshafts was used but only 27 of these cars were built. Cozette and Centric superchargers were used on some Meadows- and Gough-engined 'Nashes, all these cars being built very much to customer specification.

Body changes had included the adoption of lower lines minus running boards in 1928, and of downswept chassis in 1934; the most popular style was the 'TT Replica' two-seater. Production of all Frazer Nash cars from 1923 onwards was only 348 units, and from 1934 AFN Ltd–controlled by the Aldington family–became BMW importer for the UK. Few 'Nashes were built postwar and today AFN are British Porsche importers.

Lagonda M45

1935, Great Britain

The Lagonda company was founded in Staines, Middlesex, by Wilbur Gunn–an American from Springfield, Ohio. He had come to England in 1897 and the following year completed his first air-cooled motor-cycle in a greenhouse at Staines.

After one of his motor-cycles had won the 1905 London-Edinburgh Trial, Gunn was encouraged to build a 20hp four-cylinder four-wheeled car, which he announced in 1907. In 1909 a Lagonda light car was introduced which survived in modified form into the mid-twenties.

In 1925 Arthur Davidson designed a 1954cc ohv sporting engine for Lagonda. From this 14/60 was developed the two-litre Speed Model for 1928, when the company's parallel $2\frac{1}{2}$-litre six-cylinder was enlarged to three litres.

Unfortunately Lagondas had the reputation for being underpowered and overweight. At the 1933 London Motor Show the 1104cc Rapier with twin overhead camshafts was introduced alongside the $4\frac{1}{2}$-litre Meadows-engined M45. Now Lagonda had genuinely high performance models in their catalogue, and the racing M45R won Le Mans. Unfortunately the company was in trouble, being saved by businessman Alan Good who appointed W. O. Bentley as Chief Engineer. He produced the LG45 luxury model using the M45R engine and chassis, the LG6 of 1938 and the V12 model–considered to be Bentley's finest design.

HRG 1½-litre

1936, Great Britain

The famous 'Hurg' was the brainchild of H. R. Godfrey, who had formerly been responsible for the GN Cyclecars in partnership with Archie Frazer-Nash. Godfrey had been running his own business since this partnership dissolved in 1923, and in 1935 he developed his new car in collaboration with E. A. Halford and Guy A. Robins, choosing the title from a combination of their initials – *H*alford-*R*obins-*G*odfrey. HRG Engineering was based at Tolworth, near Kingston-upon-Thames, in Surrey.

The HRG was powered by a 1½-litre four-cylinder Meadows 4ED engine, which developed 60 bhp at 4500 rpm and was installed in a sketchy sporting frame with light-alloy two-seat bodywork. Suspension was by quarter-elliptic leaf springs at the front and semi-elliptics at the rear. In general form the car was very reminiscent of Frazer Nash practice but was lighter, had conventional transmission, could achieve 83 mph and was relatively inexpensive at only £395.

In 1939 the faithful Meadows engine was replaced by a 1½-litre overhead-camshaft Singer unit and HRG also offered a little 1100 cc sister car using the Singer Nine engine. HRG modified the engines considerably, adopting stronger valve-springs, high-lift cam profiles and free-flow induction and exhaust manifolding; in competitions the little cars shone particularly brightly in off-road Trials, since the crew's weight was concentrated well over the back axle, aiding traction immensely. HRG cars were also widely raced and rallied; 1½-litre models even tackled the Le Mans 24-Hours classic in the years 1937–39; in 1938 they finished second in their class, and in 1939 they actually won their class.

Only 36 cars had been built when war broke out in September 1939. Postwar HRGs appeared little changed from immediate pre-war trim, and they went on to even greater success in rallies, trials and minor racing events. HRG ceased car production in 1955 and the general engineering concern closed down finally in 1966. No more than 240 HRGs of all types were ever made.

'Hurg' on home ground: a 1937/8 1½-litre HRG in the Lawrence Cup Trial.

Jaguar SS100

1938, Great Britain

Ian Appleyard gave Jaguar some of their greatest postwar rally successes in this specially-built SS100 3½-litre.

The Swallow coachbuilding company, owned and run by William Lyons, first made its name with aerodynamic motor-cycle sidecars. It then progressed to a very attractive series of special car bodies. The logical sequel to the special body programme was the production of a car, and in 1931 the Coventry works indeed turned out a car using proprietary mechanical components.

This 'SS1' was powered by an untuned six-cylinder Standard 16 engine of 2054cc mounted in an all-new underslung chassis frame built to Lyons's order by Standard. All this was clad in an exciting close-coupled coupé body, with very long bonnet and distinctly rakish, sporty lines. The price was a mere £310 for a car which looked incredibly exotic for that time. A baby sister model, known as the 'SS2' and using the four-cylinder Standard Little Nine engine, was also available. In 1933 a four-seat SS1 open tourer version was introduced and a team of these cars contested the Alpine Trial. In March 1935 the SS90 two-seater sports made its debut and engineers William Heynes and Harry Weslake developed for it an overhead-valve cylinder head with twin carburettors which raised the Standard 2.6-litre engine's power output from 84 to 104bhp.

For 1936 a new range of four-door saloons was introduced, known as the 'SS Jaguar', together with a new 2.6-litre two-seater, known as the 'SS100'. The SS100 was claimed to be capable of 100mph and was wide and low, with flared front wings and a folding windscreen. With its humble pushrod engine of saloon car derivation, it was regarded with disdain by many enthusiasts as a showy 'special'. But for 1938 the SS100 was offered with a full 3½-litre engine delivering 125bhp, and was now genuinely capable of 100mph on any worthwhile stretch of straight road. These cars did very well in mid-level competition in Britain and Europe and laid the basis of the Jaguar sports-car activities which mushroomed postwar.

BMW 328
1937, Germany

The sleekly flared tail of the BMW sports car raised much comment at the time; note the disc-type spare wheel and massive fuel-filler neck.

Early in their career, the Bayerische Motoren Werke built Austin Sevens under licence in Germany, and in 1930 they released a sports version known as the Dixi-Ihle. This was BMW's first sports model, and carried the sloped-back radiator grille which was to become familiar. BMW's own touring cars developed through better-handling and more powerful designs as the thirties progressed, until in 1934 a sports version of the Type 315 1490cc six-cylinder was released, clad in a pretty two-seat body with spatted rear wheels. With three carburettors this model produced around 40bhp and was good for 78mph, and from it BMW developed the Type 319/1 1911cc 'six' with 55bhp and 90mph maximum.

The six-cylinder engine proved eminently suitable for modification, tuning and enlargement and the BMW Type 326 four-door saloon used it in 1971cc form. In the BMW 328 sports car, introduced in 1937, this enlarged 1971cc engine was mated to a tubular Type 319/1 chassis frame, and a modified cylinder head was adopted in which a hemispherical combustion chamber for each cylinder housed inclined valves operated via cross-pushrods from a single overhead camshaft. The power unit was thus endowed with the performance characteristics of a true twin-overhead-camshaft engine without the difficulties of building a twin-cam down to a realistic production price.

Fritz Fiedler was the designer responsible for the 328 and its predecessors, and his six-cylinder engine was renowned at the time for performing 'with a crisp suddenness that put it in the same class as the best overhead-camshaft engines'. It was also a very reliable unit which its post-war adoptive parents exploited fully. The BMW 328 chassis meanwhile featured precise rack-and-pinion steering; the live rear axle was suspended on semi-elliptic leaf springs, and the independent front suspension was by transverse leaf and wishbones. The result was that the car set standards of roadholding, cornering and

driving precision which had seldom been known before, certainly never in Germany.

Of more significance worldwide was the way the 328 pioneered weight-reduction and streamlining in a genuine production sports car. The car was able to produce exuberant performance from a basically humble specification, and sell at a price most could afford. Few enthusiasts, it has been said, were ever more justifiably fanatical than those who experienced the 328 in its heyday – a powerful, controllable and versatile design which has proved ageless.

At Le Mans in 1939 a BMW 328 took second place in the Index competition, and in 1940 special-bodied 328s placed first and third in the confined Mille Miglia road race in northern – and still neutral – Italy. Indeed, BMW sports cars were raced widely throughout Europe and in Britain, but their career was interrupted by World War II. After the cessation of hostilities in 1945 BMW's plants were split between the East and West German zones but BMW expertise reappeared in the Veritas, AFM and EMW sports-racing and Formula 2 cars, and in the British Bristols.

Tony Hutchings shone in sixties VSCC events with his 328. Here at Oulton Park in the rain in 1966 he heads an ERA.

BMW 328 showing off its sleek lines on the starting grid for a VSCC race at Silverstone in the seventies.

Cisitalia 'Farina' Coupé

1947, France

Piero Dusio was an Italian industrialist, sometime racing driver and car enthusiast, with the ambition of recreating Italy's pre-war motoring dominance in the postwar years. He envisaged a motor-racing circus of identical low-price racing cars, based on Fiat parts, travelling the world and showing a spectacle-starved public the pick of the greatest racing drivers battling against one another on even terms. He employed Ing. Dante Giacosa to design his Cisitalia D46 racing cars with Fiat 1100cc pushrod overhead-valve engines in advanced space-frame chassis, and before an International Formula 2 class was established in Europe postwar, the Cisitalia circus did very well. Over 50 of these tiny 800lb, 65bhp racing cars were built.

Dusio then turned his attention to greater things. The Fiat engine promised much more as the basis of a production sports car, and if he was going to build a sports car Dusio wanted the best possible body mounted on it.

The chassis for the Cisitalia sports car was derived from that of the little D46 *vetturetta*, formed as a multi-tubular spaceframe on aircraft lines. Pinin Farina produced the enveloping bodyshell without the separate wings and running boards conventional pre-war, and the result was a profoundly influential Grand Touring and sports car which set the trend of Italian high-performance-car coachwork design which remains with us to this day. The combination of low weight and low aerodynamic drag from such a clean bodyshell enabled unheard-of performance to be extracted from that humble Fiat engine, Cisitalia's 50bhp Special Sport achieving 100mph and the 60bhp Mille Miglia screaming towards 120mph and beyond.

The 1947 Cisitalia's styling features were copied on both sides of the Atlantic; but by 1948 Dusio's immense investments in his Porsche-designed Grand Prix car and other business miscalculations led to Cisitalia's collapse and Dusio's own departure for Argentina.

MG TC

1948, Great Britain

In 1937 MG at Abingdon had adopted an engine developed from the Morris Ten-Four tourer, this being a 1292cc pushrod overhead-valve unit with little really to recommend it. It was fitted in what became known as the TA model; in 1939 the TB replaced it, using a new 1250cc engine which was to prove highly successful after World War II.

In November 1945 MG introduced their TC model, retaining the 1250cc 55bhp engine, but in a wider body with a number of minor specification improvements. Top speed of the TC was quoted as 78mph and with its separate flared wings, its running boards and separate headlamps it looked everything that a sports car – at that time – should.

The TC took the British sports-car tradition into the American export market with immense success. Its good handling compensated for a poor aerodynamic shape and relatively low power – it ran out of steam towards 80mph.

Known also as the Midget, following the long MG tradition with that name, the TC had no peer for several years in the late forties. If one happened to be a sports-car enthusiast, one had to have a TC – it was the thing to do, and many aspiring racing and rally drivers on both sides of the Atlantic cut their teeth on TCs.

In January 1950 the MG TC was replaced by the TD, which offered coil-spring independent front suspension; wire wheels were replaced by discs, front and rear bumpers were standardized and again the body was widened to give a little extra cockpit room and comfort. The TD continued the TC's success despite cries of disgust from enthusiasts for the 'hair-shirt sports-car image'. In 1953 the TD was replaced by the TF model, which had its radiator raked back, a sloped slab tank at the rear and lamps fared into its wings. Extra power was available in the TF 1500 but

against the envelope-bodied aerodynamic cars from Austin-Healey and Triumph this was too little, too late. The brick-like MGs had had their day.

Veritas Saturn

1949, Germany

When peace settled over Europe in 1945 the landscape in general and industry in particular had been sorely ravaged. German factories were flattened and buried under heaps of charred rubble. Yet within a year enthusiasts were beginning to organize events for surviving touring and sporting cars–mostly hillclimbs. The BMW 328 was a popular weapon at this time and when some small establishments began building new cars, the old BMWs were a popular source of parts.

The best-known of all these postwar German products was Ernst Loof's Veritas built from 1948 to 1949 in Messkirch, from 1949 to 1950 in Rastatt and until 1953 at the Nürburgring circuit. Loof took the 1971 cc BMW 328 six-cylinder engine with which he had been involved pre-war and installed it in a tubular frame clad in aluminium coachwork of very aerodynamic open-cockpit sports-racing-car design. This competition Comet model had a 140 bhp engine and was good for 135 mph, while two more modest versions were also marketed, the Scorpion convertible and the Saturn fixed-head coupé.

In 1950 Veritas adopted a new single-overhead-camshaft 1988 cc engine built for the company by Heinkel, of aviation fame, and this engine was also offered in numerous tune-states, from 100 to 140 bhp. Five-speed transmissions were available with column gearchange on the coupés and a remote floor-shift in the competition models.

Between 1948 and 1950, when German enthusiasts were still barred from International competition outside their own frontiers, Veritas cars were entered in all manner of minor German sprints and races. They proved extremely successful at circuits such as the revived Nürburgring, Hockenheimring and Solitude, and also did well on the classical hillclimb courses as at Rossfeld and Freiburg.

The flowing curves of the XK120, one of William Lyons's classic creations. Demand surpassed all expectations.

Jaguar XK120

1950, Great Britain

In 1946 William Lyons's company in Coventry recommenced production of its SS Jaguar models, which continued to offer incredibly up-market looks and performance for very low prices. At the Brussels Salon in 1948 a streamlined Belgian bodyshell was shown on a pre-war SS100 chassis, and later that same year the London Motor Show was dominated by the appearance of the very latest, XK-series Jaguars. Two models were released, one the XK100 with a four-cylinder two-litre engine, while the sister XK120 used a six-cylinder, 3.4-litre engine. This seven-main-bearing design, with twin overhead camshafts, delivered 160 bhp at 5200 rpm from its 3442 cc. Front suspension was independent by

◄*In 1949 Jaguar test driver Len 'Soapy' Sutton drove this carefully prepared XK120 roadster – theoretically in standard mechanical trim – to average 132 mph on Belgium's Jabbeke Highway near Ostend.*

►*Another record version of the XK120, this time with bubble canopy enclosing the cockpit and 170 mph at Jabbeke!*

Jaguar family, with the classically styled SS on the right, Appleyard's famous XK120 amidships and the D-Type on the left.

wishbones and torsion bars, and the Lyons-styled bodyshell was a graceful and attractive study in flowing lines with enclosing fenders and spatted rear wheels.

Demand for the larger-engined model was so great that the XK100 never entered production. The XK120 took the sports-car world by storm. Lyons had originally envisaged a production run of just 200 to sharpen the public's appetite for the XK-engined Mark VII Jaguar Saloon due in the coming year. Tooling for full production of the XK120 delayed progress, and the first export cars did not leave Britain's shores until the summer of 1949, and the first home market cars only became available in any quantity in 1951.

The XK120 name had been chosen to indicate that the car could achieve 120mph. In fact an independent road test by *The Motor* magazine recorded a maximum speed of 125 mph, and a works-prepared model with a streamlined undershield was officially observed to achieve 132.6mph on the Jabbeke Highway record strip near Ostend in Belgium. At Silverstone in 1949 XK120s finished 1-2-3 (incidentally, painted red, white and blue); later that year Ian Appleyard won the Alpine Rally in an XK120 and Stirling Moss the Tourist Trophy. In 1950 XKs raced at Le Mans, one running third for 21 hours.

In 1951 a fixed-head-coupé version was introduced for those who wanted good looks and performance without the irritations of soft-top motoring, and in 1953 a drophead coupé added more civilization to the range.

In 1951 'Competition', or C-Type, versions of the XK120 were developed for Le Mans using bodies of revised aerodynamic form designed by the brilliant Malcolm Sayer, and an all-new tubular chassis frame. The C-Types won the 24-Hours, and came in 1-2-4 in the Tourist Trophy. Jaguar competition achievements were to include a total of five Le Mans victories during this decade.

In 1954 the XK120 was replaced by the XK140, which looked very similar but had an engine re-tuned to deliver 190bhp. In 1957 the XK150 coupés replaced the 140s, the D-Type competition-based XK-SS being intended to act as the open roadster model of the range. The omission was made good in 1958 by the release of the XK150S roadster, which used a 250bhp 3.8-litre version of the XK twin-cam six-cylinder engine and had disc brakes all round. It remained in production until superceded by the immortal E-Type in 1961.

Le Mans Talbot-Lago

1950, France

Major Antoine Lago took over the ailing French Talbot company at Suresnes in 1935, introducing a new series of six-cylinder cars with overhead valves. They were generally pleasant, if sobre in character. The four-litre 23CV model with seven-main-bearing crankshaft and – in Lago Special trim – cross-pushrod valve gear *à la* BMW delivered 165 bhp and would exceed 100 mph on any worthwhile straight. Similar cars were placed 1-2-3 in the 1937 French Grand Prix for sports cars at Montlhéry. For the 1938–39 Grand Prix Formula which catered for supercharged three-litre and unsupercharged 4½-litre racing engines, Lago authorized two lines of approach. One was a supercharged three-litre V16 engine designed by former Fiat and Sunbeam-Talbot-Darracq engineer Walter Becchia, and the second a more modest, unblown 4½-litre 'six' derived from existing power units. The V16 was rapidly shelved owing to the unavailability of sufficient capital before the war, while the unblown 4½ could not compete with the German might of that time.

After the war the situation was rather different, for against the supercharged 1½-litre Italian Grand Prix cars the 4½-litre Lago-Talbots – or 'Talbot-Lago' cars as they were actually lettered – stood a better chance, particularly since they could run through a Grand Prix distance non-stop while the highly supercharged predominantly Italian opposition had to make at least one and sometimes two stops to take on extra fuel.

Louis Chiron and Louis Rosier both won Grand Prix races with these cars in the late-forties and in 1950 Rosier and his son shared their 'Lago Record' Grand Prix model in the Le Mans 24-Hour endurance classic, the car being rigged up with cycle-type wings and head- and tail-lamps for the occasion. This remarkable special out-ran all the opposition and survived the day-long grind with imperious reliability and ease, Rosier Sr driving 23½-hours and only allowing his son a brief 30-minute stint by way of relief. This was a sensational achievement for any true Grand Prix car, quite apart from being a remarkable personal feat of high-speed driving skill and endurance.

Healey Silverstone

1951, Great Britain

Before World War II, Donald Healey was Triumph's engineering director and a very successful rally driver. When the war ended in Europe in May 1945, with ex-Alfa Romeo engineer A. C. Sampietro he set about design of a sports car to carry his own name. Their first product was offered as a well-streamlined four-seat roadster or a two-door saloon, using

a Riley 2.4-litre overhead-valve, four-cylinder engine. In 1946 the Healey proved itself Britain's fastest production saloon, achieving 104.7 mph under test. Early Healeys such as this won their class in the 1947 and 1948 Alpine Trials, and won the Touring Car Class in the 1949 Mille Miglia.

The Healeys were, however, very expensive in their home market and for 1950 Donald Healey introduced his much lighter and simpler Silverstone sports model which, while no faster than the sobre-bodied saloon, was very reliable, inexpensive and much more fun. Many budding British and American racing drivers cut their competition teeth in Silverstone models, many of them actually on the former airfield racing circuit in Northamptonshire after which the model was named.

In December 1949, talks had begun between Healey and George W. Mason, President of the Nash-Kelvinator Corporation of America, concerning the possibility of collaboration on a sports car. A prototype was run at Le Mans in 1950 which beat all American-engined opposition there. This experimental Nash-Healey finished fourth overall and in the autumn of 1950 Nash-Healeys made their production debut. Eventually Donald Healey's Warwick factory dropped the Silverstone and other models to concentrate upon Nash-Healey production for the US market – the Silverstone chassis being used virtually as was to accommodate the long-stroke six-cylinder Nash engine.

But in 1952 Healey displayed his prototype envelope-bodied Healey '100' at the Earls Court Motor Show in London. Austin were smitten with it and agreed to put it into volume production as the 'Austin-Healey', and so all other projects took second place; Austin-Healey had been born and was to dominate.

The Allard J2, a combination of British and American engineering, was as virile as it looked.

Allard J2

1952, Great Britain

Sidney Allard was an inveterate special-builder, and a sprint and hillclimb driver of considerable skill. He built his first Allard car in the mid-thirties using Ford engine and chassis, and Bugatti body and steering. His successful, mud-plugging Allards were much sought-after in trials and hillclimb circles before World War II. Immediately after the war Sid Allard returned to the hills with a fearsome device using a Steyr V8 air-cooled military engine and, eventually, four-wheel drive. In 1947 and 1948 Allard was among the top three drivers in the RAC British Hillclimb Championship and he won it in 1949.

He had formed his own Allard Motor Company in London in 1946 and introduced two fast touring cars, the models K and L, and began development of a smaller, lightweight competition car, powered by an American Mercury engine. In 1949 the Allard J2 was unveiled – an ugly, purposeful, virile competition and road-going sports car of undiluted savagery. Sid Allard intended it to carry low-price, economical and powerful American V8 power from Ford or Cadillac. With split-axle independent front suspension and transverse leaf-spring rear end, the Allard-Cadillac became a much sought-after imported sports car in the USA.

Allard-Cadillacs scored in road races right across the USA. Enthusiasts such as Erwin Goldschmidt, Tom Cole, Fred Wacker and Tom Carstens drove and entered J2s with immense success. Goldschmidt's V8 engine developed just under 300 bhp and his Allard accelerated from 0–60 mph in under seven seconds, 0–100 mph in under 14 seconds and could reach nearly 160 mph in a fearless straight line, staggering figures for the time. Sid Allard took a J2 into third place in the 1950 Le Mans 24-Hours race. Gradually, more modern concepts overtook the tail-heavy design of the J2, bred from trials and hillclimb competition, and so Allards became more sophisticated. However, they never lost the win-or-bust image characteristic of those fierce, V8-engined J2s.

Porsche 356

1951, Germany

Old Dr Ferdinand Porsche of pre-war Löhner, Austro-Daimler, Steyr, Mercedes-Benz, Auto Union and Volkswagen design fame was interned by the French authorities at the end of World War II, while in Austria his son 'Ferry' established the Porsche Büro design office at Gmünd. There he built prototypes and the first handful of production models under the Porsche name, these being lightweight sporting cars based on VW mechanical parts.

In 1950 serious Porsche production commenced at Stuttgart-Zuffenhausen in Germany. The first model was known as the Type 356. It used a rear-mounted air-cooled VW engine of 1086cc mounted in the car's tail actually behind the rear axle line and delivering around 40bhp. The Porsche 356 series developed from this root ran virtually unchanged in basic shape until as recently as 1964, when the 2000GS model was running a 1996cc Porsche flat-four air-cooled engine which delivered an honest 130bhp with performance to match.

Porsche's early reputation was built rapidly on racing, sprinting, hillclimb and rallying success, not only in Europe but around the world. The factory entered cars at Le Mans as early as 1951 when they won the 1100cc class; in 1952 Porsches won the 1500cc class in the Mille Miglia, the 1100cc class again at Le Mans, and were 1-3-4 in the gruelling open-road race known as the Liège-Rome-Liège Rally.

The stubby little Porsches handled very well with suitable suspension tuning to overcome the inherent tail-wagging effect created by their rear-engine layout; and their light weight combined with adequately powerful engines gave them a very fine power-to-weight ratio which matched their high-speed cornering abilities. In addition Porsche's engineering skills were of the highest standard; the cars had reliability built into them, a tradition still evident today in the company's painstaking test and development programmes, and almost flawless attention to quality finish and fine detail.

The Porsche 356 in Coupé, Roadster and Cabriolet forms was very neat and pretty.

Aston Martin DB2

1952, Great Britain

The well-established but always precarious Aston Martin company underwent change of ownership (by no means the first) in 1947, with its acquisition by industrialist David Brown. Claude Hill, Aston's Chief Engineer, already had a new postwar design on the stocks when this management change took place and his car was introduced as the DB1, taking David Brown's initials.

The DB1 used a four-cylinder pushrod overhead-valve engine of 1970 cc which developed around 90 bhp. It had independent front suspension and was available both as a drophead coupé and as a sports tourer. In 1949 a 2580 cc twin-overhead-camshaft six-cylinder Lagonda engine was mounted experimentally in a DB1 chassis frame and in the spring of 1950 this hybrid model entered production as the Aston Martin DB2, employing an all-enveloping two-door, two-seat coupé body. The engine was offered in standard form with a 6.5:1 compression ratio and 105 bhp, and in tuned 'Vantage' trim with 8.16:1 compression and 125 bhp.

The DB2 was a potent competition car with tremendous potential, and the works team campaigned the model very actively across Europe. In 1950 and 1951 the DB2s did well at Le Mans while never having the sheer power necessary to challenge Ferrari and Jaguar seriously, and a DB2 won its class in the 1951 Mille Miglia race around Italy.

In 1954 an enlarged version was introduced, known as the 'DB2/4' and incorporating two occasional seats in the back of the cabin for use by children or by the 'legless midgets' so beloved of contemporary motoring journalists.

Ex-Auto Union engineer Dr Robert Eberan von Eberhorst was retained by Brown to develop a true competition model for the company, and this emerged first as the DB3 — employing a tubular chassis frame and open roadster bodywork — which won the Goodwood Nine-Hours race in 1952. Late that year DB3s were fitted with 2912 cc engines which then found their way into an improved DB2/4 capable of 125 mph. In 1953 the curvaceous DB3S sports-racer was developed from the DB3 — to become the centrepiece of the beautiful Aston Martin competition line.

Peter Collins brakes his works-entered Aston Martin DB3S on the Dundrod RAC TT course in Ulster, 1955.

Austin-Healey 100

1952, Great Britain

Austin fell for the envelope-bodied good looks of the 2.6-litre Austin-powered Healey sports roadster at the 1952 London Motor Show, and a mass-production agreement was quickly forged between the two companies. This was the first real indication of postwar interest in high-performance cars on the part of a really large British manufacturer, and the Austin-Healey carved itself tremendous markets at home and abroad – notably in America at the expense of the now dated-looking, separate-fendered MGs.

The simple 2.6-litre four-cylinder Austin engine delivered about 90 bhp at 4000 rpm and, with a simple box-section frame beneath that pretty body, and independent front suspension built from proprietary components, production was relatively inexpensive. Thanks to its good shape the Austin-Healey offered very lively acceleration up to a maximum of about 105 mph and in 1953 early long-distance record-breaking attempts in search of publicity yielded 123 mph over 12 hours, and then 103 mph over 30 hours running in what was claimed to be standard trim.

The Austin engine proved itself very responsive to competition tuning and Austin-Healeys excelled in amateur speed events of all kinds. The 100M was introduced in 1954 with 110 bhp engine and lightweight bodywork, and the 100S was offered 'for export only' with 132 bhp, disc brakes and much-modified cylinder head and camshaft.

In 1956 production of the four-cylinder Austin power unit was discontinued and so the first six-cylinder Austin-Healey appeared with a slightly reduced displacement but 12 extra horsepower. The Austin-Healey 100-6 had added smoothness and torque from the six-cylinder engine and acquired a beautiful exhaust note which was music to Healey enthusiasts' ears. Weight increased, unfortunately, from around $18\frac{1}{2}$ cwt to over a ton and performance was not noticeably improved by comparison with four-cylinder trim.

The effect of modern racing regulations upon road sports cars is often hideous – here a Healey 100 has suffered.

Pegaso Z102B

1952, Spain

In the late thirties, after Vittorio Jano had left to join Lancia, design responsibility at Alfa Romeo devolved upon a Spaniard named Wilfredo Ricart. He designed many extraordinarily complex and ambitious chassis and engines before the war years, which he spent with the company, finally returning to his native Spain in the late forties. There he became chief engineer of the ENASA Pegaso company which was built upon the remnants of the old Hispano-Suiza concern in Barcelona, and which was to be involved primarily with the construction of heavy lorries.

As a prestige exercise, Ricart convinced his board of directors that a very high-priced exotic sports car would prove immensely valuable to the company. At the same time its construction would give the company trainees proper pride in workmanship before they embarked on the fairly humdrum business of building trucks and buses.

Ricart was given the go-ahead, and in 1951 his new car created a sensation at the Paris Salon. The car used a four-cam V8 engine which was virtually a Grand Prix racing design with little concession to production economies or restrictions. This $2\frac{1}{2}$-litre engine was capable of delivering 165 bhp at 6500 rpm in standard tune. It was mated to a five-speed gearbox in unit with the rear axle at the back of the car, and De Dion rear suspension coupled with coil-and-wishbone independent front suspension added still more to its Grand Prix character.

The car employed a simple platform chassis and upon this base the leading coachbuilders of the age were invited to weave their magic. Since there was no indigenous Spanish components industry to speak of, Pegaso had to build more of their own car than any similar manufacturer, even making their own brakes and bearings. Saoutchik of Paris and Carrozzeria Touring of Milan were the most popular Pegaso coachbuilders, and their products attracted astronomical prices for the time. But the cars handled and performed well and one competition version became briefly the World's fastest production car – being officially timed at 152 mph through the flying kilometre. Then Jaguar spoiled things by reaching 172 mph. The last Pegaso car was delivered in 1958.

The Spanish Pegaso was named after Pegasus, the flying horse, and had distinctly equine styling to match. Alternative bodies were happily available.

Triumph TR2

1953, Great Britain

Perhaps the Triumph TR range of inexpensive sports cars, economical to run, yet with undoubted high performance, reached its peak with the introduction of the muscular TR4 model in September 1962. By that time over 80,000 TR3 and TR3As had been produced and the new model, with its square-cut, chunky Italian styling by Michelotti, proved a worthy replacement.

It used the trusty Triumph four-cylinder engine in 2138 cc form, developing 105 bhp gross at 4600 rpm, which was transmitted via a very pleasant-to-operate, four-speed-and-reverse, all-synchromesh gearbox. The independent front suspension with coil springs was retained from the model's immediate predecessor, the TR3A, along with the live rear axle on half-elliptic leaf-springs. The TR4 was based upon a slightly wider chassis frame than that used in the TR3 family, and proved very attractive upon its introduction, reasonably priced at £1095 in open form and £1146 as a closed two-seat coupé. A detachable 'Surrey top' proved an unusual innovation, the roof panel alone detaching to leave a fixed rear screen still in place. Wind-up side windows completed the cockpit protection.

The new model had a 0–50 mph acceleration time of just over seven seconds, which was quite brisk in those days; it could cover the standing-start quarter mile in 17.5 seconds, and had a top speed of about 110 mph.

In January 1965 the TR4A was announced, this being the first all-independently-suspended sports car in the TR range. It used the Triumph 2000 saloon-car independent rear suspension with trailing-arm and coil springs; with 104 bhp it was slightly heavier than the TR4 which it replaced, 40,300 TR4s having been produced. The TR4A was never quite a match for the TR4 either in a straight line or around corners, but it rode more comfortably and was more refined. Some 28,500 TR4As were built before the model's disappearance in August 1968, the fuel-injection TR5 and restyled TR6 continuing the tradition until the advent of the modern TR7 coupé in the seventies.

▲*Enthusiasts were quick to take the Triumph TRs into competition—here a TR2 is tackling the 1953 RAC Rally of Great Britain.*

◄*TR chase on another rally night stage, 1953. . .*

AC Ace

1954, Great Britain

The famous AutoCarrier company was founded by engineer John Weller in 1908, taking residence at Thames Ditton, Surrey three years later. In 1913 the company brought out a 10 hp four-cylinder light car, and in 1919 Weller introduced what was to become his famous 1991 cc single-overhead-camshaft wet-liner six-cylinder AC engine. This entered production in 1922 and stayed in the catalogue until 1963; over the years development punched its original output of around 35 bhp up to more than 100 bhp. In the fifties this unit was to power the AC Ace shown here.

AC was under the direction of famous racing driver and entrepreneur Selwyn Francis Edge from 1921 to 1929 and cars were raced, sprinted and hillclimbed widely with great success. The company suffered in the Depression years and was acquired by the Hurlock family in 1930. In 1947 production recommenced after the war with a range of high-performance saloon and three-wheeled invalid cars, and the Petite 350 cc motor-cycle-engined convertible. Then in 1953 the Hurlocks took up a twin-tube-chassis, all-independently suspended sports-car design by John Tojeiro, wearing a body modelled after that of the successful Ferrari 166 Barchetta sports-racing car. This was fitted with the famous AC six-cylinder engine and going into production as the AC Ace in 1954 it became a terrific success; a true 100 mph two-seat sports car in classical mould with the modern lines of a wheel-enclosing bodyshell. The Ace handled very well and achieved a long string of race successes in amateur events as well as proving itself a fine road-going sports car.

AC later adopted more powerful six-cylinder Bristol engines of 2 litres' and 2.2 litres' capacity and derived from the BMW 328; eventually they turned to the economy, reliability and adequate power of the 2.6-litre Dagenham Ford engine. Disc brakes became standard on the cars in 1960 and it was upon the pretty little Ace that Carroll Shelby based his Ford V8-engined conversion – the fearsome Shelby American Cobra.

Ted Whiteaway/Jack Turner drove this AC Ace home in an astonishing seventh place overall behind four Ferraris and the victorious Aston Martins at Le Mans in 1959.

Jaguar D-Type

1954, Great Britain

While over 50 C-Type Jaguars were manufactured in series from August 1952 to 1954 for sale to private owners with competition aspirations, the Coventry works team required a new, more aerodynamic, and lighter weapon for their own front-line competition against the might of Ferrari, Maserati, Lancia and Mercedes-Benz.

For 1954 the D-Type emerged, based on a stressed-skin monocoque central chassis nacelle built on aviation principles. In this the panels curving round to form the visible external body of the car were fixed permanently to internal bulkheads, thus forming the major load-bearing structural members of the frame without resort to separate tube or girder members. A front subframe accepted the independent front suspension while a live axle was retained at the rear. This was generally maintained throughout the D-Types' long competition career and in fact it limited their effectiveness to ultra-smooth high-speed circuits such as Le Mans, which Sir William Lyons rightly identified as Jaguar's primary target – the most prestigious sports-car race in the world, victory in which would pay the biggest dividends in terms of sales.

Jaguar D-Type cars followed upon the C-Type's success; a D-Type was placed second at Le Mans in 1954, running a 3.4-litre engine against the 4.9-litre V12 engine of the winning Ferrari; D-Types won the classic endurance grind in 1955, 1956 and 1957 – Jaguar thus equalling Bentley's record of five 24-Hour race victories at Le Mans in an overall period of seven years. D-Types also won the Sebring 12-Hours race in Florida, USA, in 1955 and at Watkins Glen, USA, in 1955–6–7. About 50 competition D-Types were built, and a road-equipped sister model, the XK-SS, was laid down for full series production at the Browns Lane works. Tragically the factory was destroyed by fire in February 1957 when only 16 of the 144 mph XK-SS models had been completed, and its production was abandoned after the works was rebuilt, Jaguar concentrating on development of the new monocoque-chassis E-Type instead.

Mike Hawthorn completes a miserable race in the rain at Le Mans with his D-Type Jaguar after being involved in the third-hour tragedy which killed over eighty spectators. . .

Closely related to its Grand Prix sister, the 300 SLR resorted to air braking!

Mercedes-Benz 300 SLR

1955, Germany

When the Daimler-Benz AG directors authorized an all-out Grand Prix racing programme for 1954, they added a sports-car project to run in parallel. The new car was to use a three-litre version of the 2.5-litre straight-eight fuel-injected Grand Prix engine. The Mercedes publicity people recalled the famous gull-wing-doored, Le Mans-winning 300SL coupés of 1952, and dubbed their new sports-racing model the 300SLR. To run a mere three-litre sports-racer against the 3.4s from Jaguar and the 4.0s from Ferrari seemed like folly but Mercedes were using a pure-bred racing engine, instead of a modified production model.

The startling 300SLRs' first race was the Italian Mille Miglia in March 1955, and Stirling Moss and his navigator, Denis Jenkinson, won at the stunning average of 97.95 mph over the 1000 miles on public roads. Tested after the race, their car's engine still delivered 296 bhp at 7400 rpm, at which speed it was geared to drive the car at 170 mph. Fangio's sister car, driven solo, was second. In May Fangio, Moss and Karl Kling were placed 1-2-4 in the Eifelrennen at Nürburgring, and at Le Mans a terrific 'Grand Prix' race raged furiously during the first two hours between the 300SLRs, Castellotti's Ferrari and Hawthorn's D-Type Jaguar. This was resolved tragically at the time of the first scheduled refuelling stops as veteran French driver 'Pierre Levegh' (Bouillon) crashed his 300SLR off the tail of Lance Macklin's Austin-Healey while passing between the pit row and the crowded grandstands and hurtled into the crowd, killing himself and over 80 spectators. Daimler-Benz withdrew their cars after leading until midnight.

Fangio and Moss finished 1-2 in the Swedish Grand Prix and Moss won the Tourist Trophy from an inspired Hawthorn in his Jaguar; other 300SLRs finished second and third once the D-Type had broken. Now for the Sports Car world title, and entries were made for the Sicilian Targa Florio in October. Moss and Peter Collins shared 300SLR/0004 and brought it home to victory despite two accidents en route. Other 300SLRs were placed second and fourth. Mercedes-Benz were double Champions of the World and retired from racing forthwith.

Gullwing showing why—the pet name for the road-going Mercedes-Benz 300SL Coupé was most descriptive.

Sunbeam Alpine

1955, Great Britain

Sunbeam developed from John Marston's tinplate and japanware company in Wolverhampton, Staffordshire, which had been making bicycles since 1887 and which built its first belt-drive prototype car in 1899. The first cars to enter production were the diamond-pattern four-wheel Sunbeam-Mabley voiturettes of 1901–04, which were claimed to be skid-proof. The company was reformed as the Sunbeam Motor Car Co. Ltd in 1905 and from 1909 enjoyed a period of great success in the forefront of international motor-racing competition, bursting with prestige and dynamism and building some very fine cars indeed. In 1920 Sunbeam amalgamated with Talbot and Darracq to form the Anglo-French S-T-D combine which was to collapse in 1935, whereupon the Sunbeam company was acquired by Rootes.

The Rootes Group revived the Sunbeam name after World War II and in 1953 it was given to the Alpine, a two-seat version of the 2.3-litre Sunbeam-Talbot 90, with its overhead-valve, four-cylinder engine. The Alpine won four coveted Coupes des Alpes in the Alpine Rally of 1953, and Stirling Moss won a Gold Cup and Sheila van Damm the Ladies' Cup in the Alpine Rally of 1954. The Alpine was a genuine 100 mph car when properly tuned and was very attractive and handsome for its day, although very much a convertible touring car rather than a true sports roadster.

The model was discontinued in October 1955, but the Sunbeam Alpine name was revived in 1960 for a low-slung unit-construction sports roadster with 1494 cc 78 bhp engine which was handsome in a typically understated Rootes manner, and was good for a genuine 98 mph. This model was a civilized and comfortable road tourer; it was too heavy for serious competition although the factory team campaigned lightened versions in the longer sports-car events such as Le Mans and Sebring. From this model the Sunbeam Tiger was later developed, with its American Ford V8 engine.

MGA

1956, Great Britain

As early as 1951 the MG technical team at Abingdon, headed by the brilliant Sydney Enever, realized that the company would head for disaster if Nuffield Group management insisted upon retaining in production the T series cars with their outdated, non-aerodynamic body shapes. For Le Mans in 1951 a prototype MG was constructed carrying a streamlined wheel-enclosing bodyshell on a TC chassis frame. This was intended as an experimental production prototype, and Enever decided it required a new chassis frame which indeed made it into an eminently saleable proposition, but the remote management controlling MG's destinies at this time did not recognize the new car's potential.

In 1952 the new Triumph and Austin-Healey sports cars appeared with their enveloping bodies, and hammered more nails into the coffin of the T-series MG. Finally the go-ahead was given for Enever's brainchild to enter production. It finally emerged in Sept-

ember 1955 as the svelte and beautiful MGA, after a team of three prototype cars had run with some success at Le Mans.

With a 68 bhp engine, the MGA matched the 98 mph maximum speed of a race-tuned 97 bhp MG TF with unstreamlined bodywork. The better aerodynamic shape of the MGA clearly paid tremendous dividends and using the 1489 cc four-cylinder engine shared by the ZA Magnette saloon the MGA was set fair for considerable commercial success. In fact its production run in seven years was to exceed 100,000.

A rather exotic twin-overhead-camshaft engine was also developed for the MGA, but this version was produced in relatively very modest numbers, and was not to prove very popular, although effective in competition use. A Twin-Cam MGA was the first MG to have front disc brakes as standard, and in 1960 discs appeared in the 1588 cc MGA 1600, which replaced the smaller-engined original model. Both two-seat open roadster and fixed-head coupé models were available, the MGA combining beautifully responsive and safe handling characteristics with considerable agility and speed.

The pretty MGA Hardtop in an unlikely setting – grass-track autocross at Taunton in the sixties.

Chevrolet Corvette

1956, United States

General Motors' first Corvette was thought of as nothing more than a glittering show car, all chromium plate and promises. Not even GM, it was said, could possibly market a special-bodied sporting device like that for under $4000. But Chevrolet knew better. The Detroit giant astounded the American enthusiast public by entering full production with the car and at last doing something, just a little, to fend off the imported European sports-car threat.

The first Corvette appeared at the end of 1953 using a stock six-cylinder 3.7-litre side-valve workhorse engine tuned to deliver 160 bhp. The two-seat roadster bodyshell was the world's first in relatively large production to employ glass-reinforced polyester techniques; it was a 'glass-fibre' moulding. The Corvette offered the American market adequate performance combined with reasonably flat cornering – despite the softness of the suspension, which made for a much more comfortable and willowy ride than that offered by the hard-sprung Europeans, such as the MGs. Of the European cars, it was said that if one drove over a silver dollar lying in the road, one could tell whether it was heads or tails. Although their engine displacement was actually modest compared with some of the imports, independent road test reports soon confirmed that the earliest Corvettes were quick enough, easy to drive without tiring out their driver, and furthermore that they could out-drag Jaguars from the traffic lights while being able to out-corner any other American car on the open highway.

The first hand-made production Corvette leaves the Flint (Michigan) production line, June 30, 1953.

The completely restyled 1958 Hardtop model with V8 engine and deeply scalloped sides.

▲ *Moody public relations and advertising photography does its best for the 1956 Chevvy Corvette, which was actually a much larger motor-car than it appears to be from this angle.*

◀ *When GM adopted glass-reinforced-plastic body construction for the Chevrolet Corvette it created a sensation. Here this 1953 release photograph shows the major mouldings pre-assembly.*

Corvette '73 featured the later-style body with integral urethane-covered bumper up front.

Even a pussyfoot driver could feel content and safe at 108 mph along the 'straightaways' in his plastic-bodied special, and some Corvettes just happened to go together right and they were good for 110 mph at least. Where the Corvette lost marks was under braking, for hard driving in hilly country would fade out its drum brakes completely. The provision of Powerglide automatic transmission as standard in the early cars disappointed many would be owners but as the type developed so its sporting essence was distilled and refined.

In 1956 a 4.3-litre overhead-valve V8 180 bhp engine was introduced in the Corvette line; by 1960 handling had been greatly improved and a fuel-injected 290 bhp engine option was available.

For 1963 the styling was completely revised and the redesigned Corvette Stingray appeared with multi-curvature body and pointed nose with retractable podded headlamps, all powered by a 5.3-litre 360 bhp V8 engine. This machine proved immensely popular on the American market and appealed to one or two exhibitionists in Great Britain and Europe. In 1968 the model was again restyled with engine options of 5.3 litres and seven litres.

In the course of the fifties and sixties, the Corvette became America's best-loved indigenous sporting car, and one which in its many variations has an immense following today in Corvette Clubs throughout America.

Jaguar E-Type

1961, Great Britain

When the E-Type Jaguar was unveiled as a replacement for the XK150 series in the spring of 1961 it was greeted with acclaim from the whole motoring world. William Lyons's rare eye for graceful styling had produced a svelte design which brought into reality almost every enthusiast's dream of the ultimate sports car.

Jaguar had developed their D-Type monocoque centre-section theme in an interim prototype sports-racing car which had raced in Briggs Cunningham's white and blue American colours at Le Mans in 1960. The E-Type now used the 3781 cc XK twin-overhead-camshaft six-cylinder engine in 265 bhp trim, and carried all-independent suspension, at last replacing the live rear axle of the Cs, Ds, and 120s, 140s and 150s. Disc brakes also appeared all round; at the rear they were located inboard on the cheeks of the final-drive unit. Open roadster and coupé body-styles were available; the fixed-head coupé was the fastest model on independent road test – though carefully prepared by the Coventry works – achieving a genuine 151 mph. Such performance was available for the remarkably low price of £2196 but insurance companies reacted by heavily loading drivers they thought too young, too inexperienced, or simply suspect.

The E-Type became *the* sports car of the sixties, exemplifying all that was best in British high-performance car engineering, although the early 3.8-litre models were always handicapped by their slow, heavy old Moss four-speed gearbox. This was replaced in 1965 by a superb all-synchromesh gearbox in conjunction with an enlarged 4235 cc engine, and in 1966 the 2+2 Coupé was introduced with raised and rather ungainly roofline to accommodate occasional rear-seat passengers. In 1970 another new model was introduced, based on a long 2+2 wheelbase; with 5.3-litre V12 power, this was the ultimate E-Type Jaguar, although the clean-cut styling of the original models had long since been despoiled by American Federal Safety Regulation requirements.

Not a lot of space, but bags of grace and pace typified the Jaguar E-Type, particularly the early 3.8-litre roadsters as seen here racing at Brands Hatch.

Lotus Elite

1961, Great Britain

Colin Chapman's Lotus Engineering company grew from a part-time hobby building trials specials in the late forties and early fifties into a company which dominated the competition world, thanks to Chapman's advanced thinking on chassis and suspension design. His aerodynamic-bodied, ultra-lightweight sports-racing cars set new standards during the fifties, and in the middle of the decade Chapman began to plan production of a road-going flagship model to capitalize upon his prestige and competition success.

This model, the Lotus Type 14 Elite, made its debut in 1957 and was not only startlingly attractive but also of absorbing technical interest. It was the first Lotus intended primarily for road use, and yet naturally it retained immense competition potential. It was based upon a fully stressed monocoque body/chassis unit which was unique in being constructed entirely of glass-fibre mouldings; the sophisticated all-independent suspension was bolted onto strong-points moulded into the structure. For structural reasons the car was only offered as a fixed-head coupé. Power was provided by a 1216cc Coventry Climax single-overhead-camshaft engine delivering about 75bhp and this endowed the good-looking and very lightweight projectile with a standard top speed of some 112mph. Disc brakes were fitted all round.

Unfortunately the Elite proved a pretty deceiver, certainly of its manufacturer, for production costs were far higher than anticipated and minimal profit was shown on each one built; very soon Lotus was selling the cars at a loss. The unique plastic body structure proved vulnerable to fatigue cracks and many customers became disgruntled, but on the racing circuits of the world the Lotus Elite proved itself a formidable contender in the 1300cc class, winning its class at Nürburgring in 1959, and at Le Mans every year from 1959 to 1964.

Today the better-maintained surviving Lotus Elites are much-prized collectors' cars and they are still raced with tremendous success at amateur level in Britain, Europe, Australia, the Far East and America.

All Lotus cars bear the unmistakeable stamp of the firm's founder, Anthony Colin Bruce Chapman, and the Lotus badge carries his initials.

Triumph TR4

1962, Great Britain

Perhaps the Triumph TR range of inexpensive sports cars, economical to run, yet with undoubted high performance, reached its peak with the introduction of the muscular TR4 model in September 1962. By that time over 80,000 TR3 and TR3As had been produced and the new model, with its square-cut, chunky Italian styling by Michelotti, proved a worthy replacement.

It used the trusty Triumph four-cylinder engine in 2138cc form, developing 105bhp gross at 4600rpm, which was transmitted via a very pleasant-to-operate, four-speed-and-reverse, all-synchromesh gearbox. The independent front suspension with coil springs was retained from the model's immediate predecessor, the TR3A, along with the live rear axle on half-elliptic leaf-springs. The TR4 was based upon a slightly wider chassis frame than that used in the TR3 family, and proved very attractive upon its introduction, reasonably priced at £1095 in open form and £1146 as a closed two-seat coupé. A detachable 'Surrey top' proved an unusual innovation, the roof panel alone detaching to leave a fixed rear screen still in place. Wind-up side windows completed the cockpit protection.

The new model had a 0–50mph acceleration time of just over seven seconds, which was quite brisk in those days; it could cover the standing-start quarter mile in 17.5 seconds, and had a top speed of about 110mph.

In January 1965 the TR4A was announced, this being the first all-independently-suspended sports car in the TR range. It used the Triumph 2000 saloon-car independent rear suspension with trailing-arm and coil springs; with 104bhp it was slightly heavier than the TR4 which it replaced, 40,300 TR4s having been produced. The TR4A was never quite a match for the TR4 either in a straight line or around corners, but it rode more comfortably and was more refined. Some 28,500 TR4As were built before the model's disappearance in August 1968, the fuel-injection TR5 and restyled TR6 continuing the tradition until the advent of the modern TR7 coupé in the seventies.

Austin-Healey 3000

1963, Great Britain

When the 2.6-litre Austin-Healey 100–6 was announced in 1956, it was obvious that its sales potential could be greatly enhanced by the competition possibilities of the six-cylinder engine. However, the bodyshell used by the 100–6 was more a 2 + 2 than a traditional two-seat roadster, offering shelf space behind the front seats for two small children or a 'transverse normal adult'. This made the car some 450lb heavier than the basic four-cylinder Austin-Healey 100, but weight was trimmed with great success by the British Motor Corporation competitions department at Abingdon, who took the 100–6 and modified it not for racing, but for the arduous sport of rallying.

Thus the 'Big Healey' came into existence in rally trim, finished in the factory colours of red and white, and festooned with long-range driving lamps and spotlights. The raucous six-cylinder cars came into their own after 1959 when the engine was taken out to 2912cc to produce the Austin-Healey 3000. By 1965 this power unit was delivering a reliable 160bhp, while the body and chassis proved robust enough to withstand the worst buffets and batterings of the roughest international rally conditions. Essentially the rally Healeys were excellent high-speed open-road competition cars, but when driven in the early sixties (notably by dirt-road-bred Finnish aces like Timo Makinen) they also made their mark on loose-surface forest stages in such gruelling events as the British RAC Rally.

In 1961 and 1962, crewed notably by Stirling's sister Pat Moss with Ann Wisdom, and by the Morley twins Donald and Earle, the big Healeys won the testing Liège-Rome-Liège, and in 1964 the Austrian Alpine Rally. The production road-going versions basked in the reflected glory of these achievements until February 1968, by which time over 50,000 of the large-engined Austin-Healeys had been built.

Sadly, management panicked at the introduction of restrictive legislation in the large American export market, killing off the Healeys in favour of the misbegotten three-litre-engined MGC. Datsun in Japan studied the Healey theme, and updated it in their very successful 240Z, showing what could have been done with just a little enterprise.

▲ *At its best when sitting down under power, the Austin-Healey 3000 rally car in the rain on one Epreuve.*

► *The legend of the 'Big Healey' grew in the forest rallies of the late fifties and sixties in Britain and on the open road rally races of the Continent. Here in the forests that barking three-litre six-cylinder scatters loose stones as a works driver powers out of a curve.*

MGB
1963, Great Britain

Sydney Enever's technical team at Abingdon developed their first unitary-construction monocoque sports car for release in 1962, replacing the separate-chassis MGA. This model, known logically enough as the 'MGB', did not win itself instant affection from the MG test-drivers. Whereas the MGA had been a forgiving and good-handling car right from prototype the Abingdon company had teething troubles with the MGB in making the chassis structure sufficiently rigid to offer predictable handling qualities and adequate controllability. After a fairly fraught development period the car was licked into shape, and the 1795 cc 95 bhp MGB was released to the public in 1962 – offered as a two-seat roadster – and in October 1965 a fixed-roof MGB GT coupé joined the range.

The MGB was a handsome-looking car with an appeal quite different from that of the flowingly styled, almost glamorous little MGA. It remained in production with few apparent changes other than those dictated by new safety regulations in its major markets until 1980, when MG in its existing form was killed off by its parent, the British Leyland group. By this time the MGB was old-fashioned and obsolescent, even though market demand was still very high.

Certainly the MGB once properly developed formed a solid, road-going sports car with good handling, and many competition variants appeared which did quite well up to the late-sixties in minor long-distance sports-car races and the occasional rally.

The car sold very well in the USA until in cut-price deference to new safety regulations vast impact-absorbing bumpers had to be added to the existing bodyshells and, more serious, they had to be located at a regulation minimum height. To achieve this level, the export MGB version was jacked up on its suspension into the air; in this form, further hindered by smog-limiting emission-control devices which strangled the engine's already rather modest power output, the poor MGB not only did not like going round corners, but also had barely sufficient power to step smartly along the straights.

Shelby-American Cobra
1964, United States/Great Britain

Large-engined sports cars have always had a special appeal for American enthusiast drivers, and in the early sixties Carroll Shelby (formerly with the Aston Martin works team and winner at Le Mans in 1959) decided that he was going to build just such a car. He looked around Europe for a suitable vehicle for a cheap-to-buy, easy-to-maintain yet very powerful, proprietary American V8 engine, and he chose the AC Ace chassis and body being built at Thames Ditton.

In 1962 the first Shelby American prototype cars were constructed, mating a 4.2-litre Ford V8 from Detroit with the British twin-tube chassis and pert Ferrari Barchetta-derived bodyshell. Initial experience of these Anglo-American hybrids showed-up – predictably – handling and strength problems which were rapidly remedied, although the cars always remained demanding machines to drive really quickly.

After 75 of these Cobra sports cars had been built, the 4.2-litre Ford engine was replaced in 1963 by the latest 4.7-litre variant which offered 195 bhp. In England the cars were known as AC Cobras, while in the USA Shelby American took full credit with no mention at all of the AC parentage.

By 1965 Shelby was shoe-horning the latest 6989 cc Ford V8 engine into further-improved Cobra shells. These were more bulbous, with extended wheel-arch eyebrows

▼ *Shelby American intended their Cobra to be 'all bear' and succeeded.*

to enclose ever-wider wheels and tyres, and with suspensions uprated to accommodate the seven-litre engine's greater mass and power output, which hovered around the 345 bhp mark. Acceleration and braking in a straight line was shattering, 0-100-0-100-0 mph being achieved in less than 30 seconds! Identified by their 289- and 427 cubic-inch capacities respectively, the 4.7- and seven-litre Cobras were literally earth-shaking devices, although in competition around corners drivers claimed they had to shut their eyes and hope. Even the smaller car could punch itself from rest to 60 mph in 5.5 seconds, and reach 138 mph flat-out. A special Daytona coupé competition model with enveloping bodywork was constructed with a 380 bhp 4.7-litre engine – this was capable of 195 mph. Construction of all models ceased in 1968.

Exhausting drive – Dan Gurney's Cobra in the 1964 Targa Florio.

Morgan

1965, Great Britain

The remarkable Morgan Motor Company has survived into the eighties with an ever-faithful band of enthusiasts queuing up to buy its sports cars, which are still constructed in the traditional manner. Their ash-framed bodies feature separate flowing wings and running boards, and offer fairly spartan accommodation. Morgans even today have a 'thirties' character they are still 'the real thing', and not by any means a replica.

Under the direction of Peter Morgan, son of 'H.F.S.' the founder, the Malvern Link company developed during the fifties, building first a Plus Four model fitted with the tuned 70 hp Standard Vanguard engine, and later, from 1954, a 90 bhp Triumph TR2 engine which provided genuine 100 mph performance for the first time.

In 1955 a small Morgan was reintroduced, fitting into the range below the Plus Four and using the faithful 1172 cc side-valve Ford four-cylinder engine which had powered the

Opposite lock—this Morgan Plus Eight was an optimistic entry in the 1980 Monte Carlo Rally, practising here at Shelsley Walsh.

old F4 Morgan three-wheeler, discontinued in 1950. This new four-wheel model was known as the 4/4 Series 2; later the overhead-valve Ford 105E engine was substituted as Dagenham dropped the side-valve.

Meanwhile the Plus Four kept pace with Triumph TR engine development. A streamlined coupé model very similar in general appearance to the Lotus Elite was offered in 1964 to a roar of disapproval from the Morgan traditionalists, and from 1968 into the eighties to date Morgan have used the quiet, reliable and excellent Rover $3\frac{1}{2}$-litre V8 engine in their Plus-Eight model, capable of a comfortable 125 mph on its large wheels and tyres without losing any of the age-old Morgan character.

In many ways Morgan is an anachronism in a modern motoring world – but Thank God for Independence and the retention of old standards of workmanship and pride.

Ford Mustang

1966, United States

The Ford Motor Company's answer to Chevrolet's early success with the Corvette was the Thunderbird, introduced in 1955. This handsome if not exactly sporting machine was larger and more powerful than the Corvette, with 4.8 litres, 200 bhp and 115 mph. However, instead of continuing the two-seat roadster theme, Ford backed out of the market in 1958, modified the Thunderbird range to carry four seats and so killed it as a 'sports car'.

At the United States Grand Prix meeting at Watkins Glen in 1962, Ford's engineers displayed a two-seat mid-engined roadster exercise which they called the 'Mustang'. In very different form, the Mustang emerged in production in 1963 as a compact front-engined four-seater offered with a massive range of optional equipment. This allowed the successful young man or woman to personalize the car to choice, making it as sporty or as civilized as they wished. The Mustang was available with six- and eight-cylinder engines in four sizes from 2781 cc to 4728 cc and with power output ranging from 101 bhp to 210 bhp. The Mustang in these various forms was good for anything from 95 mph to a full-blooded 120 mph; and when Carroll Shelby adopted the model and 'breathed upon it', he produced the Shelby Mustang GT350 with modified bodywork, engine, suspension, brakes, fittings and trim. The road variant gave 306 bhp and the competition model 350 bhp from the 4.7-litre V8 engine, Ford's famous '289' cubic-inch unit. In 1967 Shelby introduced the GT500 with the full seven-litre V8 giving 360 bhp in road tune or 400 bhp for the race track. Even the road car could top 135 mph and still a vast range of options was available to normal owners who would never dream of hazarding their production Mustangs in competition.

The Ford Mustang featured strongly in competition saloon car racing in Britain, Europe and the USA in the mid-sixties, but in truth it was essentially a modern 'sporting', rather than 'sports', car.

Works Ford rally Mustangs slamming past the Reims pits during a sixties Tour de France Automobile.

The Ferrari Dino is a classic in the mid-engined tradition, V6 engine and Pininfarina bodywork being combined with superb road-holding.

Ferrari Dino 246GT

1970, Italy

Enzo and Laura Ferrari's only son was christened Alfredo after his uncle, Enzo's brother, who had died of natural causes during World War I. In Italian 'Alfredino' is the affectionate form of this christian name, and the diminutive of Alfredino is simply 'Dino'. Poor Dino Ferrari died from complications of progressive muscular dystrophy in 1956. Within months of his death a new 1500cc V6 Formula 2 engine was running on the Maranello test-beds, and Mr Ferrari decreed that this engine should carry the name Dino after his beloved son; in 1958 Ferrari Dino V6 Grand Prix cars carried Mike Hawthorn to his World Drivers' Championship title.

In 1959 Dino V6 engines powered sports and Formula 1 cars and by 1961 Ferrari V6s were in a position to win the World Championship again. During the sixties a string of developing V6 and V8 engines carried the Dino name, and come 1967 and the introduction of a new 1600cc Formula 2, Ferrari wanted access to a production engine from which a racing Formula 2 unit could be developed. He sold Fiat the idea of their building

Cromodora cast-alloy road wheel for the Ferrari Dino.

Pininfarina's reverse-curved rear screen was a fascinating styling feature of the Ferrari Dino family's beautiful bodywork.

a production version of his two-litre V6 so that he could 'modify it for racing' and so slide what was effectively a purebred 1600 cc racing engine around the new Formula 2 rules. He was successful, although the Formula 2 cars were not, until the end of 1968.

At that time the road-going small Ferrari, using the two-litre Dino V6 engine, was introduced as an exotic but inexpensive car wearing pretty Pininfarina bodywork. This model was later updated to accommodate a full 2.4-litre 246GT V6 engine, and in spider and coupé forms these Dinos proved mouth-watering road cars offering tremendous performance and superb road-holding. The V6 engine was set transversely amidships behind the two-seat cockpit, and the little cars were good for 140 mph with acceleration to match. Their steel bodies have proved subject to rust and decay, which is a tragedy, for they are truly glorious little machines and the seventies' supreme embodiment of the sports-car tradition. They were discontinued in 1973, and replaced by three-litre transverse V8-engined Dino models.

Dino was run as a marque separate from Ferrari, until production outran that of their larger brethren. This was an interesting point for Ferrari dealers and importers around the world considered that to create a 'Dino' marque was itself a marketing error, for their customers required true Ferraris and were put off by the idea of buying something which was regarded by owners of the more expensive and larger models as 'not a proper Ferrari'. In truth, however, the Dino name had achieved enough in competition to be truly honourable.

In fact it was the Dino cars which introduced truly modern chassis engineering into the Ferrari tradition, for while the larger, predominantly V12-engined Ferraris had superb engines and gearboxes their handling and cornering behaviour was often 'antique' at best and 'truck-like' at worst. With the svelte little Dino range Ferrari really offered adequate, indeed outstanding road-holding in the modern idiom, and that development made an immense difference to this most classical of post-war classic manufacturers.

Lamborghini Miura

1972, Italy

During the sixties the Italian tractor manufacturer Ferruccio Lamborghini took to car manufacture more or less as a hobby. He pursued his enthusiasm for matchless cars and created a line of luxurious, very high-performance Grand Touring cars which achieved the highest standards of technical sophistication.

The first Lamborghini car appeared at the Turin Show of 1963. It was a front-engined GT coupé using a $3\frac{1}{2}$-litre four-cam V 12 engine with six Weber carburetters. This very Ferrari-like device looked more like a true racing than a production road-going unit, the whole project being supervised by ex-Ferrari engineers Giotto Bizzarini and Gian Paolo Dall'ara.

Italian exoticar – Lamborghini Espada.

After unforeseen problems in preparing the Lamborghini 350GT for production it began to appear in numbers in the mid-sixties. Meanwhile Dall'ara had produced an even more exciting design in the Miura – named after the pugnacious Spanish breed of fighting bull, Lamborghini's response to Enzo Ferrari's Prancing Horse.

This strikingly unconventional Miura packed a four-litre V12 engine set transversely behind the two-seat cockpit amidships, driving through spur gears to the clutch and five-speed transaxle assembly. This remarkable assemblage produced and handled no less than 420 bhp at 8000 rpm, and yet with simply incredible reliability and lack of temperament the car could poodle around city streets or streak along the Autostrada del Sole at anything up to a genuine 180 mph. It was the World's fastest production car for some time.

In the seventies the Miura's crown was inherited by the remarkably uncompromising Lamborghini Countach of generally similar specification, which takes its name from a Modenese slang word meaning more or less 'Corr . . .'; and with reason.

Lola T212

1972, Great Britain

Former architect Eric Broadley began his competition-car construction company in the late fifties more or less as a hobby. He had built himself a club-special which he named 'Lolita', and the 'Lola' cars which followed were built on the precept that whatever Lola wants, Lola gets.

During the sixties Lola Cars Ltd worked briefly with the mighty Ford Motor Company of Dearborn, Michigan, to develop the original GT40 model, and that programme really established Broadley's company in fit form to 'go it alone'. Currently – in 1979–80 – Lola Cars of Huntingdon are the world's largest manufacturer of pure competition cars, building sports and single-seater models for virtually every International racing class.

The European two-litre Championship came to prominence in the sports car world in 1970 and remained there for five years during which time Lola products featured strongly. Broadley's Lola T212 design, for example, featured a central monocoque chassis which carried its Cosworth-Ford four-cylinder FVC engine of 1790 cc (and yet 245 bhp!) in a tubular cradle behind the cockpit. All-independent suspension featured, as was conventional competition practice, with coil-spring/damper units and wishbone location. Gearbox was a proprietary Hewland five-speed transaxle. These cars gained considerable success in Britain and western Europe while many were sold to the USA for club racing there. In 1972 the Lola T290 version appeared with cleaned-up, sharper-nosed body lines; the Cosworth BDA or FVC engines were taken out to as much as 1930 cc and produced over 250 bhp at 9500 rpm. In 1973 this model was followed by the T292 with a large wing over the tail and lengthened nose, with 275 bhp Cosworth power. The 1974 model was a further improved car known as the T294, while in 1975 completely revised aerodynamic thinking produced the T390 with curved and bulbous lines reminiscent of an Alpine-Renault.

Throughout all these years the two-litre Lolas were amongst the best-selling of a modern sports-racing line, directly in the tradition of the Lotus, Jaguar, Aston Martin and MG racers of the fifties and before.

Alpine-Renault A110

1973, France

Automobiles Alpine was founded at Dieppe by Jean Redélé in 1955. He was an enthusiastic Renault dealer who that year built a competition-orientated special using Renault components. The car promptly featured successfully in the small-car class of the classic Italian Mille Miglia race. He used moulded glassfibre bodywork for his cars, offering them as two and four-seat coupés.

In 1961 Redélé offered two new body styles, one a 2 + 2 GT coupé and the other a very aerodynamic two-seat Berlinette. The glassfibre bodywork of the latter was mounted upon a central-backbone chassis which carried the one-litre Renault engine in various tune-states and forms at its extreme tail, behind the rear axle line. Redélé called the Berlinette the A110 'Tour de France' after the great speed competition for GT cars still being held in France and immediately neighbouring countries each autumn.

The A110 actually set out with only 850 cc and less than 50 bhp to propel it, but Redélé became increasingly confident as success followed success, Renault showed increasing interest, and their engine consultant Amédée Gordini produced ever more effective competition power units for the Dieppe factory's use. Renault 8, then Renault 16, and finally very highly-tuned 1800 cc power units developing close on 200 bhp were slotted into the sleek little competition coupé's tail. The chassis were well-balanced, extremely agile, and responded magnificently when asked to change direction or dodge suddenly. They were also far, far stronger than they looked, and these attributes rendered them magnificently suitable for International rallying, in which the Alpines' extraordinary agility gave them the edge over often far more powerful rivals. In 1973 the works Alpine-Renault team from Dieppe won the prestigious World Rally Championships.

The A110 was finally removed from the catalogues in 1977, to be replaced by the less attractive-looking A310 model. The A110 was a modern classic sports car by any standards, and sadly missed.

TVR Vixen

1973, Great Britain

The Lancastrian specialist sports-car company, TVR, derives its name from that of its originator, Trevor Wilkinson, who began building tube chassis and later complete car kits in the late 1950s. His tiny Blackpool works produced a two-seat coupé known as the Grantura, with a glass-fibre body, powered by Coventry Climax or Ford 100E engines and suspended on Volkswagen components. In 1960 the

Mark II Grantura retained the same basic body shape–which has proved timeless–but adopted Ford 105E and MGA additional power units. John Turner devised a new Mark III tubular chassis frame with double-wishbone suspension in 1962 and these TVRs from Blackpool really began to make a name for themselves in minor-level club races and rallies.

Meanwhile the TVR company itself had undergone many management and ownership changes, but the little Blackpool-built cars soldiered through good times and bad with adequate success. A Ford V8-engined Anglo-American model known as the TVR Griffith was produced in conjunction with an American promoter, and after his company collapsed TVR continued V8 production.

Trevor Fiore designed a graceful and very Italianate new body style in 1965, which was later developed as a separate (and unsuccessful) marque using the large Austin-Healey chassis and running gear. In 1966 TVR produced their Mark IV with 1800 MGB power unit and in 1968 a move was made down-market with the little Tina, using Fiore bodystyling and a Hillman Imp engine. The 1800S model was also offered with Ford Cortina power–which was a very practical proposition–and they called this model the TVR Vixen.

Come the seventies and a Triumph-engined 2500M appeared, fitting a new multi-tubular backbone chassis frame under that so-familiar but still not dated body style. In 1972 this chassis was applied to all TVR engine options and later in the decade a convertible appeared, together with a Turbo version, using a highly tuned, turbocharged Ford V8 power unit.

Triumph Spitfire Mk IV

1973, Great Britain

The Triumph company found its way into the sports-car markets of the world with the TR series. There appeared also to be a gap in the market for an inexpensive, small-capacity sporting car, easy to drive and not too fast–just the thing in fact for the youthful or lady enthusiast who liked the sporting idea if not the full-blown practice.

In October 1962 Triumph released the Spitfire 4, an 1147 cc two-seat model developed from the successful Herald saloon with its backbone chassis. Styling was by the Italian house of Michelotti; mechanically there was little difference from the Herald except for the use of softer springs to accommodate the lightweight body. A twin-SU-carburettor version of the Herald 1200 engine was used, producing 63 bhp at 5740 rpm. Rack-and-pinion steering gave the minute turning circle of only 24 feet and the four-speed gearbox with synchomesh on all but first gear was another attraction. This little car was good for 93 mph, would reach 50 mph from rest in around 12 seconds and its fuel consumption was a major selling point at 35 mpg. It was priced at £729 and sold very well.

From September 1963 overdrive and a hardtop were available for the now best-selling little Spitfire, and in March 1965 an improved Mark II Spitfire offered a 67 bhp engine, improved trim and furnishing, and was altogether a less spartan device. Fastback coupé versions were raced at Le Mans with some success in the 24-Hours endurance race and in October 1966 the Spitfire-based two-litre Triumph GT6 coupé was announced. In early 1967 the Spitfire was itself uprated to house a 1296 cc engine; with 70 bhp at 6000 rpm this Mark III version offered 95 mph, and a 0–50 mph time of around 10 seconds–still with 36 mpg fuel consumption. A Mark IV, with changes in styling and mechanical detail, followed in 1970.

Datsun 240Z

1975, Japan

During the sixties the Japanese motor industry slammed into world wide export markets with shattering effect. Companies like Datsun and Toyota proved extremely adept at producing practical and reliable motor-cars to a price, and with back-up service which the average customer could not afford to ignore. The Japanese also looked at the performance-car field, and Datsun in particular scored a notable success with their 240Z coupé model.

The Kwaishinsha works had produced its first car in 1912 in Tokyo, and in 1914 a more successful second car was produced which the company named the Dat, using the initials of the proprietors, K. Den, R. Aoyama and A.

Works Datsun 240Z rally car at home in the rugged and dusty conditions of the East African Safari Rally.

Takeuchi. In 1926 Kwaishinsha merged with Jitsuyo Jidosha Seizo to form the Dat Automobile Manufacturing Company of Osaka, actually making trucks. When car production resumed in 1931 they were called Datson ('son of Dat'), but in order to suggest the national Japanese emblem of the rising sun the name assumed its final form, Datsun, in 1932.

Immediately after World War II Datsun cars were based on the American Crosley design and during the fifties the company built nothing less (nor more) than the Austin A50 under licence. It was during the sixties that Nissan Datsun modernized their range and attacked international markets so successfully while the market at home was heavily protected against imports.

Effectively the 240Z coupé was the result of examination of the Austin-Healey 3000 six-cylinder model, which had been allowed to wither on the vine in the face of stringent new US Federal Safety Regulations in 1968. Datsun developed their own 2.4-litre single-overhead-camshaft six-cylinder unit enclosed in strikingly attractive streamlined coupé bodywork; with five-speed all-synchromesh transmission, all-independent suspension, front disc brakes and 160 bhp the big Datsun carried on where the big Healey had been allowed to die. The 240Z was fast, handled adequately well, was quite economical and very reliable – very much a sports car for the seventies.

Triumph TR7

1975, Great Britain

The long line of Triumph TR sporting cars had developed as far as the long-nose TR6, when, in January 1975, the two-door two-seat coupé TR7 replaced the old traditional line with its directly traceable 'fifties' pedigree.

The new TR7 coupé was designed very much with the American market in mind, which in part excuses the exaggerated wedge-shape profile with ugly, sculptured indentation in the body sides. The wheelbase, at only 7 ft 3 in, was rather too short for comfort when combined with a 4 ft 7 in track front and rear, and the result was a car which, though nimble and quick to change direction, rode rather choppily and felt a little unstable along any bumpy straight.

It was marketed as the 'front-engined car with the mid-engined looks' but in truth combined perhaps the worst points of both designs. Its four-cylinder engine with a single overhead camshaft displaced 1998 cc and developed about 80 bhp at 5400 rpm. A four-speed manual gearbox transmitted power to the rear wheels. The bodywork was an all-steel unitary structure with independent front suspension and a live rear axle, on coil-springs all round. Disc brakes featured at the front and drums at the rear, with dual hydraulic circuits and vacuum servo assistance. This little car was claimed to have a 108 mph top speed, not half as fast as its appearance implied, but it has proved reasonably successful both in the USA and in Europe.

The basic TR7 was progressively updated by the introduction of a five-speed gearbox and later by the installation of the Rover 3500 V8 light-alloy engine which at last gave it real power. Known as the TR8 in this guise, the car was rallied extensively by a works team with some success, and in British Airways sponsorship livery the TR8's certainly looked the part as their drivers fought their way with visible excesses of power around the rally stages of the world.

Fiat X1/9

1976, Italy

During the seventies the sporting car in general developed as a two-seater coupé. American Safety Regulations at one time looked likely to ban the open-top car for good, hence Triumph's replacement of the TR6 by what was in original form a coupé-only TR7, and Jaguar's replacement of the E-Type roadster by the XJ-S coupé-only model. Some manufacturers were less cautious, and Porsche developed the so-called Targa top in which the fixed windscreen surround was augmented by a roll-over bar-cum-rear-screen surround behind the cockpit, and a rigid but detachable roof panel was provided to insert between them. This system had appeared in the Triumph TR-series 'Surrey top' but Porsche gave it its modern name following their successes in the arduous Targa Florio race around Sicily. This race favoured open-cockpit cars, which nevertheless had to retain good aerodynamic shape for the long Buonfornello Straight along the coastline.

In fact it was upon the Targa Florio circuit in Sicily that Fiat first introduced their extraordinary X1/9 Targa-top sports car to the world's motoring press. What this vast manufacturer of utility cars had done was to take the transverse engine and front-wheel-drive package from the existing and very successful Fiat 128 and install it amidships behind the two-seat cockpit of a striking, wedge-profiled bodyshell both designed and made for them by the house of Bertone, at Grugliasco outside Turin. The rigid roof panel could be removed and stowed away in the front locker and thus converted the figure-hugging little X1/9 proved

And everything opens! Fiat's X1/9 was a daring gamble which paid tremendous dividends for Fiat and stylist-cum-body-builder Bertone.

to be a staggeringly enjoyable little car.

The high-revving 1300cc four-cylinder engine could punch the car along at speeds as high as 103mph, and it would accelerate from 0–60mph in about 12.6 seconds. It covered the standing-start kilometre in 34.5 seconds, which was a very worthy achievement, returning an albeit modest 21.5–28mpg. Disc brakes on all four wheels were without the servo which in other Fiat models had proved disturbingly harsh in operation, and the X1/9 set new standards for nimble and agile handling, controllability, braking and general driving fun in the very best sports-car traditions.

In 1978 a 1.5-litre version was added to the range and as the eighties dawn the baby Fiat is still very attractive to all who enjoy their motoring.

Lotus Esprit

1979, Great Britain

The Lotus Elite ceased production in 1963, to be replaced by the more practical Elan, and later the Europa mid-engined coupé. In due course, too, a new Elite was offered, with a Vauxhall-based engine developed by Lotus.

At the Turin Motor Show in 1972 Giorgetto Giugiaro's Ital-Design styling house produced a mid-engined study which placed the new Lotus engine behind the two-seat cockpit of à sleek and very low-slung coupé of gorgeous appearance. Colin Chapman was very interested in this project, and detailed engineers Tony Rudd and Mike Kimberley to form a team to develop a production variant of this Italian study. Working closely with Giugiaro in Turin, Lotus evolved their striking Esprit two-seater which was introduced in October 1975.

The four-cylinder engine of 95.2 mm bore × 69.2 mm stroke displaces 1973cc and carries twin overhead camshafts actuating four valves per cylinder. Peak power is 160bhp at 6200 rpm and this is delivered to the rear wheels via a five-speed gearbox. The Esprit uses a backbone chassis frame with all-independent suspension front and rear based directly on Lotus' vast experience of single-seater and sports-car racing worldwide. Vacuum servo-assisted disc brakes appear all round, with fat tyres on neat, lightweight, cast-alloy wheels and very precise rack-and-pinion steering.

Unfortunately, in the Series 1 cars at least, the gearchange was not very good owing to the problems, common in mid-engined cars, of mating up a cockpit change with rear-mounted transmission while at the same time insulating the cockpit area from noise and vibration. There was body boom generated in the engine bay, and it took quite a while to sort these difficulties out to the customers' satisfaction. Finally it was achieved, with the Esprit Series 2.

In 1978 a turbocharged version of the pretty Esprit was developed by one of Lotus's major distributors, and so brought real power to a sports car worthy of the eighties.

Panther Six

1979, Great Britain

Bob Jankel established his Panther Westwinds company essentially to build what were virtual modern-day replicas of the SS100 pre-war sports car. They proved very successful, they were very well made and beautifully finished, and Jankel's company progressed to make all manner of replica-type sporting vehicles plus some high-quality special coachwork projects. This programme culminated in the series production of the Panther Lima GM-Detroit and Vauxhall-engined two-seat sports car. This was very much an evocation of the MG TC era, expressed in modern high-quality glass-fibre and based on a production-car chassis pan, engine, gearbox and running gear offering no spares and servicing problems.

Panther now wanted a flagship model for their newly broad range, and during 1976–77 Jankel was impressed by the performance on circuit of the six-wheeled Tyrrell Project 34 Grand Prix cars. He determined to make a sporting car to excel all others.

Panther installed side-by-side bench seating for three at most in the open cockpit of a mid-engined six-wheeled car of outstanding comfort and opulence. To achieve the stunning performance required Jankel chose a seven- or eight-litre American Cadillac V8 engine mated to a General Motors automatic transmission. What was more, he contracted Ak Miller – a prince of the Californian hot-rod movement – to fit twin Garrett AiResearch turbochargers to the V8 engine to boost it to 600 bhp.

The result could have been either stunning or frightening. In fact Panther adopted what they called 'softly-softly' ride control from Jaguar saloon suspension units and the Panther Six, as the car was christened, emerged as an incredibly comfortable device with tremendous power and performance potential both in a straight line and through corners. The author was the first journalist to drive it, and although underdeveloped it impressed him deeply with its bite under brakes and turning into corners. Unfortunately Panther found themselves crippled over supply of the special front tyres required, and production of the six-wheeler hung fire.

Porsche Turbo

1979, Germany

As the seventies reached their close undoubtedly the most potent sporting car generally available in the sports-car market was Porsche's Type 930 Turbo. This had evolved from the long-serving Porsche 911 family whose world-wide success in production trim was based on one of the most comprehensive and expensive competition programmes any major motor manufacturer has ever been involved in.

The turbocharged model was introduced at the Paris Salon of 1974, stunning all observers with its enormously wide tyres enclosed in massive fender cowls, its front air dam and prominent tail spoiler.

The flat-six-cylinder engine of the 911 family, as used in the Porsche Carrera model in three-litre unsupercharged form, had reached the end of its development and only supercharging promised a future for it. Porsche

Ultimate development of the 911 family: the Porsche Turbo, combining familiar looks with stunning performance.

chose to adopt exhaust-driven turbo-supercharging in which an impeller vane introduced into the exhaust stream is forced to revolve. Its spindle then drives a similar impeller in the induction side of a supercharger unit which then feeds the engine with mixture under pressure. This system had been developed by Porsche in competition from 1972; in the road-going Porsche Turbo 2 + 2 coupé the three-litre engine was boosted at a maximum of 121 lb per square inch pressure, and responded by producing an enormously reliable – and not uneconomical – 260 bhp at 5500 rpm. What was more important, the turbocharged engine gave an incredibly wide power band which allowed the Porsche Turbo driver to outperform almost anything on the public road without wearing himself out by continual resort to the gear lever. The Turbo would potter happily about at 2000 rpm in top gear, but with the driver's foot hard down it would reach 50 mph from rest in a mere 3.6 seconds, 100 mph in 11.8 seconds and 125 mph in less than 20 seconds. Top speed was around 155 mph at 6000 rpm – but the brakes in these early cars had a hard time and it was not until the advent of the 3.3-litre Turbo version in September 1977 that the model finally received the brakes which its enormous performance really demanded.

De Lorean

1980, Great Britain

John Zachary De Lorean was a chief executive of the mighty General Motors Corporation – the world's largest private enterprise – until he suddenly decided to go his own way and build his own high-performance cars. Many others had done this before, but very few had ever succeeded. De Lorean was ambitious in an unusual way and set out to build a sporting car primarily for the American market which would break new ground in automotive technology.

He wanted a mid-engined two-seater car with outstanding styling, including gull-wing doors *à la* Mercedes 300SL; it would combine advanced safety features with good performance and fuel economy. Structurally the car was to employ a number of plastic processes for which De Lorean held US licences.

In December 1974 De Lorean and his chief engineer, Bill Collins, another ex-GM man, went to the Turin Motor Show in Italy to talk with the world's top stylists and Giorgetto Giugiaro was their choice as the man with the most attractive ideas.

Giugiaro consequently styled the De Lorean prototype which was built up to completion in October 1976. A Ford V6 was contemplated initially as the power unit for this car but the first prototype was fitted with a 2.2-litre Citroen CX four-cylinder engine and transaxle, while later cars were to use the Renault 30 power-train. This featured the alloy Peugeot/Renault/Volvo V6 engine which delivers about 120 bhp at 5500 rpm. This was good enough for a 0–60 mph acceleration time of about eight seconds and fuel economy of the order of 22 mpg in city driving and 29 mpg on longer runs. With all-independent suspension and four-wheel disc brakes the De Lorean was remarkable for its stainless-steel external panels, and the company initially planned to open a factory on a disused US military base in Puerto Rico once its plans had received sufficient financial backing.

Subsequently, however, De Lorean was attracted to Ulster by British Government subsidies, and the development of his dream car was undertaken in collaboration with one of his major potential rivals, the Lotus Group. The company ceased production in 1982.

LUCAS
LUCAS
LUCAS
STP

RACING CARS

Introduction

What is a racing car? For the purpose of this volume it is not merely a car which is raced occasionally, but one which has been built specifically for competition purposes. There are a few exceptions, such as the 1890s Panhard, and the 1903 Gordon Bennett-winning Mercedes '60', which to some extent doubled as road-going non-competition designs. Most of them won the 'classic' tag by success in competition; some others were notably unsuccessful in racing but deserve mention for technical innovation or sophistication.

It was in 1894 that the first properly organized motoring competition took place, in France, in the form, virtually, of a rally. Then, in 1895, motor racing got under way.

The first true motor race in the modern sense was run on May 29, 1895 in northern Italy, from Turin to the village of Asti and back. It attracted only five contestants: a Benz car, a four-seat Daimler omnibus, a steamer christened *Staffetta* and two Hildebrand-und-Wolfmüller motor-cycles. The steamer broke down, the Benz driver abandoned all hope at Asti and Simone Federmann in the Daimler won at 9.6 mph from the two motor-cyclists. The world, or that tiny part of it which heard of the event, evidently grunted and turned over the page for some interesting news . . .

Twelve days later the age of racing began in France, in spectacular style. The newly formed Automobile Club de France had organized a race over no less than 745 miles, from Paris to Bordeaux and back. The Club received 97 entries, which yielded 22 actual starters and 9 road-weary finishers. First home, almost six hours clear of the second-place man, was Emile Levassor, at the tiller of one of his own 4 hp twin-cylinder Daimler-engined Panhard-et-Levassors. Levassor drove the whole distance single-handed, and, apart from brief halts at controls, non-stop. He had reached his anticipated crew changeover point three hours early; his relief driver and mechanic had not yet appeared there but he would not wait. Nor would he hand over the tiller to his own riding mechanic, and clattered on through the night by the dull glimmer of his oil-burning carriage lamp. After a stop at Bordeaux lasting only four minutes, Levassor turned about and charged through another night. He fell asleep once, hitting a bank without doing any severe damage to his steed, and 48 hours 48 minutes after leaving Paris he returned there to the Finish line—having averaged a remarkable 15.07 mph. By the race rules first place was awarded to the first four-seater car, and Levassor's special two-seater was denied victory, but there was no doubt of his heroic reception by the French public.

Thereafter, the development of motor-racing and the cars themselves was explosive. Engines grew larger and larger as designers sought more power and therefore speed, virtually regardless of weight. Company promoters and publicists saw the new sport as an instant way of selling their wares. The buying public neither knew nor cared that the latest racing Panhard, De Dietrich or Mors bore little resemblance to the car they would buy. Racing success, then as now, proved that the manufacturer knew what he was doing.

Slowly the character of racing changed, as the city-to-city era was ended by accidents. Circuit racing took over, usually on public roads closed for the occasion. The first Grand Prix, in 1906, was run on a 64-mile circuit, while laps above 50 miles were commonplace. These circuits had to be covered up to a dozen times to create a worthy test of speed and endurance.

From 1900 to 1905 the pre-Grand Prix international competition for the Gordon Bennett Trophy concentrated (largely unwarranted) attention upon strictly prescribed three-car teams from each interested manufacturing country. Cars changed during this period from early primitives with 5-litre 25bhp engines to 10–16-litre models developing as much as 90–120bhp at 1000–1400 rpm.

The Gordon Bennett competition's regulations took no account of the relative strengths of different national motor industries. The French became heartily tired of this, and founded their Grand Prix race in 1906. This was a one-off annual event. It was not until after World War I that other national Grand Prix races began to join the French Grand Prix to form the annual series of up to 17 World Championship Grandes Epreuves which we see today. There was a very low-key World Championship run by the governing body of the sport, the AIACR, in 1924–25 but it was not a competition which sparked the public's imagination until re-introduced in 1950 as the Drivers' World Championship.

But this was far in the future. When Grand Prix racing began, cars were derived from Gordon Bennett models with four-cylinder short-stroke engines of 12–18 litres capacity developing over 100bhp. The riding mechanic was a very vital member of the crew in those days. The driver had his work cut out fighting to aim his car over the loose-surfaced roadways of the time, and controlling the throttle and brakes; his mechanic was constantly at work maintaining air-pressure in the fuel tank (to force fuel through to the engine) and monitoring oil drip-feed to lubricate that hard-working engine.

In 1908 there was wider use of dropped frames curving down between front and rear axles to lower the overall height, and scuttles began to be built up to afford driver and mechanic greater protection from the elements.

Grand Prix racing was abandoned from 1909 to 1911 in response partly to a trade recession, but largely to Germany's defeat of the French establishment in the 1908 race. Voiturette racing – effectively the 'Formula 2' of the time – took precedence and new marques came to prominence, such as Peugeot and Delage. When the Grand Prix was revived in 1912 using tarred road surfaces for the first time, there was a titanic struggle between the new voiturette-derived lightweight cars and the old chain-drive, big-engined monsters as exemplified by Fiat. With their 7.6-litre cars Peugeot initiated the trend towards more modest engines in lighter chassis made fast

Théry changing tyres on his Brasier at the Col de la Moréno depot during the 1905 Gordon Bennett race around the Auvergne, France.

by technical sophistication, while the voiturette race run concurrently showed the form of things to come with a flock of refined three-litre cars.

Peugeot dominated the racing season of 1913 – as related in detail in later pages – and such Brooklands track-derived British trickery as streamlined wind-cheating tails was expressly forbidden by the French club's regulations.

By 1914 the Peugeot-inspired practice of using four valves per cylinder actuated by twin overhead camshafts had become generally accepted, and engine output at the new 4.5-litre limit averaged 120 bhp at 2800 rpm in cars weighing about 1.2 tons. Seat height was lowered by the use of double-drop-frame side rails and Peugeot learned the Brooklands Brooklands lesson – fitting tapering-tail bodies. During the Great War years racing was confined to the USA where 1914 Grand Prix cars like the Mercédès and Peugeot were widely successful in track racing. In 1919 the post-war straight-eight-engined Ballot cars achieved superior lap speeds despite their traditional bolster-tank bodywork in place of the fashionable tapering tail-cones, but this regression proved to be temporary.

In 1920 in Europe long tails in the American Indianapolis track-racing style became generally accepted, and when the Grand Prix was revived in 1921 based on three-litre US regulations, there was almost universal use of long-stroke straight-eight-cylinder engines with more than two valves per cylinder, giving 115–120 bhp at 3500–4200 rpm. Long-tailed bodies with staggered seating were widely used, the riding mechanic being tucked behind and to one side of his driver, often with a cut-out tail cowl into which he could fit his arm, embracing the driver's back.

When a new two-litre limit was applied for Grand Prix racing in the years 1922–25, four- and six-cylinder engines were revived, developing 80–90 bhp at 4500–5000 rpm in small cars weighing about three quarters of a ton. Road surfaces were much improved during this period and the mass start became general, superseding a staggered start for each car at 30-second intervals.

By 1925 the riding mechanic had been banned. He was thought to be vulnerable in the event of an accident, and no longer necessary, since tyre changing and mechanical repairs on shorter circuits could usually be done in the service pits. Driving-mirrors therefore became obligatory. The French Grand Prix was run for the first time on an artificial road circuit, the 7.6-mile course at

A 1932 Alfa Romeo P3 'Tipo B' Monoposto being raced in a VSCC event by its owner, Bill Summers.

the Montlhéry Autodrome. This was also the first Sunday Grand Prix.

New 1.5-litre regulations were applied in 1926, and manufacturers adopted straight-eight engines with roller bearings, offset to the left-side of the car. The offset driver's seat was mounted down alongside the propeller-shaft, which now passed through the position previously occupied by the mechanic. A very low driving position was thus achieved. Bugatti were very successful with a more traditional car, but in 1927 their dual-purpose true two-seater-type car was out-classed as the new purpose-built Grand Prix models achieved full development.

From 1928 to 1930 there was general disregard of regulations as money was short, and two-litre models from 1922 to 1925 regained popularity. There was widespread use of two-seater bodies to make cars usable in both Grand Prix and sports car form. In the years 1930 to 1933 Formule Libre (Free Formula) became accepted without restriction on size of engine or car. Works-sponsored teams and drivers reappeared to oust the amateur enthusiasts once again, and largely unsuccessful experiments were made with engines of between four and five litres capacity developing up to 300 bhp. What was decisively successful was the introduction of single central seats placed above the propeller-shaft in slim-line bodies, reminiscent of the type used in US track events during the twenties. The definitive form of single-seater racing car had evolved.

In 1934 the new 750 kg maximum-weight Formula saw the German state-backed teams of Mercedes-Benz and Auto Union developing sledge-hammers to crack the Grand Prix nut.

Giuseppe Campari winning the 1924 French Grand Prix at Lyon in his works Alfa Romeo P2.

All Grand Prix cars except Bugatti now featured a centrally located single seat, and all, with the exception of Auto Union, adopted straight-eight engines.

During the unlimited-capacity years up to 1937 Mercedes-Benz perfected the most powerful Grand Prix engine ever seen, delivering 646 bhp on the test-bed and around 600 bhp in the chassis. In 1980, 43 years later, the best-developed Formula 1 three-litre engines were capable of a genuine 500 bhp and the near-perfected 1.5-litre turbocharged Renault V6 ran nearer 550 bhp on occasion.

This titanic Formula ran into 1937 and was replaced by a new and restrictive class demanding a minimum – as opposed to the former maximum – weight limit of 850 kg for cars of only three litres supercharged or 4.5 litres unsupercharged. Central-single-seater bodyshells were mounted on chassis powered by complex V12-cylinder engines developing 400–450 bhp.

Meanwhile voiturette racing had continued alongside the Grand Prix classes, with the 1.5-litre supercharged class of the thirties dominated by ERA, Maserati and finally Alfa Romeo, with one brief intrusion by Mercedes-Benz. There were other popular classes too, for 750 cc and 1100 cc racing cars, but there is insufficient space in these pages to cover examples of these cars which were in any case derived largely from Grand Prix and 1.5-litre Voiturette practice.

World War II was to put a temporary end to motor racing. In Europe, still the centre of road-racing competition, the full stop came in 1940, while in American track racing, the calendar ran uninterrupted through the season of 1941. On both sides of the Atlantic, racing recommenced in 1946, despite shattered road communications around Europe, and extreme shortages of fuel, tyres and even sparking plugs. The pre-war Italian voiturettes were a dominant force in Europe; only Alfa Romeo entered a true works team, and succeeded accordingly.

The three years 1947 to 1949 saw competition between what had been the pre-war 1500 cc Voiturette class and the 4.5-litre unsupercharged pre-war division of Grand Prix models, and this proved a far fairer match than the three-litre supercharged class raced so expensively by the German teams of 1938–39.

In 1950–51 a successful challenge to the dominance of supercharged racing engines was mounted by Ferrari with a complex 4.5-litre unsupercharged V12 engine of modern design. With over 350 bhp in chassis weighing less than one ton, Ferrari eventually toppled Alfa Romeo, and when the great Milan factory team retired at the end of the 1951 season there was no realistic competition left to face Ferrari. Consequently in 1952–53 Grand Prix World Championship events were thrown open to two-litre unsupercharged Formula 2 cars, this Formula having grown in recent years along with the British-developed 500 cc motor-cycle-engined Formula 3, now recognized internationally.

The first true postwar Grand Prix Formula was established in the period 1954–60 with an allowance of 2.5 litres unsupercharged and only 750 cc supercharged – a comparison which finally killed off the supercharged racing engine for a quarter-century. While the 'Two-and-a-halves' contested Grands Prix racing using alcohol-based fuels and running 300-mile distances through 1957, a new 1.5-litre Formula 2 was developed which was to bring to the fore smaller manufacturers who specialized in building only racing cars – teams like Cooper and Lotus in Britain. A factor in the trend towards specialist racing-car constructors, using engines and gearboxes from outside proprietary sources, was the concentration on personality rather than marque resulting from the introduction in 1950 of the Drivers' World Championship. This offended some of the manufacturers involved, and some mass-production companies who also built racing cars drifted away towards rallying and other promotional activities.

When the 1958 regulation changes were

made, which allowed Grand Prix distances to be reduced to 300 km or two hours minimum, and also banned the use of alcohol fuels in favour of straight aviation petrol, the small British teams with their supply of proprietary engines from Coventry Climax began to come to the fore. Their small lightweight chassis far outperformed the tradition-steeped frames from Continental manufacturers such as Ferrari and Maserati, and in particular Cooper's rear-engined – or more properly 'mid-engined' – layout signposted the way ahead. By 1960 all worthwhile Grand Prix teams and many involved in the minor Formulae had their engine behind the driver's cockpit and Lotus in particular broke new ground with sophisticated suspension systems and high-quality chassis forms which are still with us today.

In 1961–65, what had been the 1.5-litre Formula 2 was elevated to Formula 1 Grand Prix status with a few extra safety regulations added, and then in 1966 the return of power was heralded by the adoption of a new three-litre unsupercharged Formula. This has come to be dominated by Cosworth-Ford with their DFV V8 three-litre engine, and more recently by Ferrari's powerful flat-12 312 unit from Italy. But in 1977 Renault returned to Grand Prix racing with the first engine of the current Formula to make use of the 1.5-litre supercharged option, in this case with an exhaust-driven turbo-supercharger. In the French Grand Prix of 1979, the French nationalized *Régie* scored the first forced-induction Grande-Epreuve win since Fangio's Alfa Romeo 159 won at Barcelona in 1951.

In the pages which follow some of the outstanding cars of motor-racing's long and varied history are covered in some detail. They trace essentially the story of Grand Prix racing with occasional diversions into minor Formulae and American track racing on the relatively few occasions when these have reversed the trend of technical development and have fed something back into the Grand Prix mainstream. The arrangement is basically chronological, with the cars in each year placed alphabetically, although there are one or two departures from that rule where it makes sense to group particular cars together, or where a particularly important car merits a two-page spread to itself. I have sought to describe within the space available the background to each car's design and development, and its significance in the overall scheme of things; there is also some reference to the personalities who drove them, and to those who worked on design and development. On occasion the selection of one car rather than another may seem arbitrary, but I have tried to give a balanced picture overall.

▲ *British Historic Car Racing class events have remained immensely popular. Here Charles Lucas hurls his Maserati 250F sideways round the now defunct Crystal Palace course in London.*

► *The Montagu Trophy, presented by the then Lord Montagu of Beaulieu as the team prize for the Gordon Bennett competition in the early 1900s.*

The 'Paris-Bordeaux' Panhard of 1895.

Panhard-et-Levassor
1895, France

The first motor race of all time took place in 1895, from Paris to Bordeaux and back. Panhard promotor Emile Levassor drove throughout that epic 48-hour journey, at the tiller of his Panhard-et-Levassor, powered by a 4 hp, two-cylinder Daimler-Phénix engine. However, during a repeat race in 1896, this time to Marseilles, Levassor lost control in swerving to avoid a stray dog, and was to die the following year reputedly from lingering head injuries.

By 1898 motor racing was under way in earnest – this was the great age of city-to-city racing in western Europe. Building on their moral victory in 1895, Panhard became the most successful marque of the period, and in 1898 their cars dominated the season. Fernand Charron won the Marseilles-Nice and Paris-Amsterdam events; Leys won the Course de Périgueux; and the great *Chevalier* René de Knyff won the Paris-Bordeaux. De Knyff was prime mover behind the Automobile Club de France and he combined his organizational talents with great skill and stamina as a racing-driver.

As you can see, these early racing machines were high-built and spindly, and often with tiller instead of wheel steering. The engine was in the front, coupled to a three-speed gearbox which drove through bevels to a countershaft set across the car with chain sprockets on each end. Chains then drove the rear axle and road wheels. The chassis was wooden, as were the wheels. The early tiller steering had contributed to Levassor's accident. Only the strength of the driver prevented it being deflected by obstacles in the car's path and in 1898 Panhard adopted steering with reduction gears which lessened this effect. They also adopted a 2.4-litre four-cylinder 8 hp engine that year, plus pneumatic – in place of solid – rubber tyres.

There was no weather protection for the driver and his riding mechanic, who wore heavy protective clothing, often with leather face-masks and mica-lensed goggles for protection against the wind, rain and flying stones. This was a sport for the strong!

Renault Voitur Légère
1902, France

Whereas the Panhard of the 1890s grew larger and larger engines in search of greater power and speed, the Renaults which raced at the start of the twentieth century provided a better pointer to the way racing cars would develop. The first Renault racing cars had appeared in the 1899 Paris-Rambouillet race with shaft-driven back axles and tiny De Dion engines. Louis Renault and his younger brother Marcel were good engineers, and they developed their cars rapidly through 1900–01, increasing engine size and adopting lateral radiators for cooling.

At the time there were effectively four classes of racing car. The French club had formalized regulations in 1901 for cycle-cars below 250 kg in weight; for voiturettes, 250–400 kg; for light cars, 400–650 kg; and for heavy cars, over 650 kg. In effect the light cars and voiturettes became virtually indistinguishable. The Renaults were typical. They ran 499 cc single-cylinder 'thumper' engines in 1900, and for 1902 Renault developed their own power-unit, a 3758 cc four-cylinder.

The 1902 Renaults failed on their début in the Circuit du Nord race, but in their second

Renault Voiture Légère, 1902.

event, the 615-mile Paris-Vienna, Marcel Renault's new four-cylinder machine led the whole field, including the vast heavyweight cars, by reason of its good chassis, better braking and nimble power. In 1903 the great age of city-to-city racing came abruptly to a stop in the Paris-Madrid race, which was abandoned following Government intervention at Bordeaux after a series of crashes which caused at least ten fatalities among competing crewmen and spectators. Louis Renault's light car averaged 90 mph on the straight tree-lined stretch from Bonneval to Chateaudun, and recorded 62.3 mph from Paris to Bordeaux. There Louis was informed that his brother Marcel had been killed together with his mechanic after overturning.

Marcel's death brought Renault racing-car development to a brief halt but the company's early-century light cars were shining examples of what could be done with modest engines installed in a light and practical chassis frame.

Mercédès 60 hp
1902, Germany

Wilhelm Maybach's Mercédès design in these early years was notable for its use of a pressed-steel chassis frame instead of wooden, selective-gate gearchange of the type still common today, honey-comb rather than gilled-tube radiator, and mechanically actuated rather than atmospheric valve gear.

The cars made their debut in racing trim at Pau in 1901, but failed dismally. The new marque redeemed itself at Nice soon after, where Werner's 5.9-litre 35 hp car covered the flying-start kilometre faster than all other petrol-engined cars, and was beaten only by Léon Serpollet's Land Speed Record Steamer. Werner won the Nice-Salon-Nice race and Mercédès were 1-2-3 at the important La Turbie hillclimb. The 1902 Mercédès competition models used a 6.8-litre engine rated at 40 hp. Although Stead's car won at La Turbie, the 1902 cars did not achieve the great success expected of them and their very advanced mechanical specification, but in 1903 the new 9.2-litre 60 hp model could achieve 75 mph on the very indifferent roads of the time. Mercédès took 3½ minutes off the La Turbie record, and fully 5.9 seconds off the standing-start-mile time along the sea-front record strip at Nice. They were to progress to great things in coming years, including the 1903 Gordon Bennet Trophy race in Ireland which fell to Jenatzy's modified touring '60'.

'Gordon Bennett' Napier
1902, Great Britain

Selwyn Francis Edge won the Gordon Bennett Trophy for Great Britain in 1902, driving a green-painted British Napier from the company which he directed and promoted so effectively in these early days.

At the time Britain was virtually nowhere in international motoring and the Gordon Bennett competition was in truth moribund when Edge scored his success. This came in the Paris-Innsbruck section of the Paris-Vienna race in which Marcel Renault performed so nobly, but the nominated French Gordon Bennett team cars all failed on the rough roads and the tough Napier was the sole surviving qualifier to reach the Austrian city.

Edge had promoted Napier through competition since 1900 when he had won a Bronze Medal in the British 1000-Miles Trial with a chain-drive 8hp twin-cylinder model. He ran a 4.9-litre 16hp chain-drive Napier in the Paris-Toulouse-Paris race of 1900, and for 1901 Montague Napier developed a monstrous 17,157cc four-cylinder car for the Gordon Bennett event only to have it debarred for using foreign-made tyres. This car was timed at around 67mph and won two classes in the straight-line hillclimb at Gaillon. It delivered some 103bhp but weighed over two tons and proved a barely practical proposition. In fact it was this car which prompted an outright racing-car weight limit of 1000kg.

For 1902 Napier realized the error of the big-engined way, and turned instead to a shaft-driven light car of only 6.5 litres rated at 30bhp but delivering closer to 45 in reality.

It was from the chosen dark green livery of the Napiers that 'British Racing Green' evolved as the United Kingdom's recognized international racing colour. While the accepted colour for British cars became green, German cars were white or later silver, French cars blue, Italian cars red, Belgian cars yellow and American cars white-and-blue.

Drivers Jarrott and Stocks with their Gordon Bennett Napiers, 1903.

Mors 'Dauphin'

1903, France

Emile Mors' cars gained immortality with one race victory – and even then it was victory on the only stage actually run of the last city-to-city classic – the 1903 Paris-Madrid which was stopped at Bordeaux. Fernand Gabriel's 70 hp Mors with its boat-prowed wind-splitting bodywork started 168th from Versailles, start order having been decided by ballot. While Louis Renault's light car was the first to reach Bordeaux, it had started third, and was able to run ahead of all others, with an open road stretching far ahead. Renault's time of 5 hrs 29 mins 39.2 secs to Bordeaux represented an average of 62.3 mph. Gabriel had to tear through choking clouds of dust and exhaust fumes, through showers of flying pebbles, past accidents and crash debris, to finish third on the road at Bordeaux, having passed 165 other participants. His time of 5 hrs 14 mins 31.2 secs represented an average speed of 65.3 mph, which subsequently won him the race overall and confirmed this drive as one of the greatest feats in motor-racing history.

The big Mors had weighed in at 995 kg compared to the Renault's 650 kg. Its four-cylinder engine had a bore – the width of one cylinder – of 119 mm and a stroke – the distance travelled by the piston as each combustion forces it down in its cylinder bore – of 165 mm, to give a displacement capacity of 11.6 litres. The engine was carried in a pressed-steel chassis frame with solid axles at front and rear, and transmission was via a four-speed gearbox and chains.

Later in 1903 the Honourable Charles Stewart Rolls (the Rolls of Rolls-Royce) drove a similar 70 hp Mors in a British speed trial at Welbeck and recorded a speed of 84.68 mph (136.28 km/h) over the kilometre. This was a measure of these cars' potential performance. After 1904, however, Mors made little further impact upon the racing scene – Gabriel's heroic performance in that tragic Paris-Madrid being left to etch their name in history.

Gabriel's fantastic Mors, depicted in the Art Nouveau Michelin Tyre building in London's Fulham Road.

Grand Prix Renault

1906, France

In 1906 the restrictive regulations of the Gordon Bennett competition were swept away in favour of a Grand Prix open to all comers and to be organized annually by the Automobile Club de France. Unlike today when there may be 16 Grand Prix races in a season, in 1906 there was only *the* Grand Prix, run once only, in June. It was not until after World War I that other national Grand Prix races joined the French event on the calendar and became regular fixtures.

Renault won that very first Grand Prix at Le Mans, running not on the famous 24-hour sports-car circuit but on a mighty 64-mile public road course laid out to the east of this industrial city. The winning car was driven by the Hungarian Ferenç or Franz Szisz ('Zizz') who had been Louis Renault's riding mechanic in earlier races and was a works test-driver at the company's Billancourt factory in Paris.

His car was a very conventional 13-litre four-cylinder, and this first victory for Renault in the heavy car class was achieved more by reliability and by steady and competent driving than by sheer speed or animal flair. In addition Renault's long adherence to the shaft-drive back axle did little to aid the car's cornering ability, for the axle itself was not very well located and around corners tended to move in an arc and so create what we now call oversteer – the tail slid wide and tried to spin round ahead of the front wheels. Szisz's driving was sufficient to cope with this shortcoming, and the introduction of detachable-rim wheels greatly reduced the time lost in slashing punctured tyres away from fixed rims. With detachable rims the whole rim complete with deflated tyre could be unbolted and replaced by a new rim ready-mounted with an inflated spare cover.

The Grand Prix was run in two parts, six laps each day on both June 26 and June 27 with the cars impounded overnight. Szisz won the first day's racing after 5 hr 45 min 30.4 sec on the road, having averaged 66.8 mph for the 384 miles; on the second day he won again, after a further 6 hr 28 min 36.6 sec on deteriorating road surfaces, to average 63 mph overall for the total 769.9-mile distance. The big Renault won from Felice Nazzaro's Italian Fiat, and Albert Clément's Clément-Bayard. Paul Baras' Brasier made fastest lap overall, on the first day, with an average of 73.3 mph.

Renault did not contest another Grand Prix race until July 1977, when their 1.5-litre turbocharged RS01 prototype car started in the British Grand Prix at Silverstone.

◄*One of the ex-GP Renaults during a 24-Hour 'grind' in America circa 1907.*

▲*Ferenc Szisz's GP Renault defending its victory in the first Grand Prix race of all time, at Dieppe, 1907.*

Grand Prix Benz

1908, Germany

Old Carl Benz considered, once he had perfected his first horseless carriage in 1884–85, that the motor-car as such had been made. He could see very little future improvement upon his invention.

In 1906 this fiercely conservative engineer left the company he had founded to form Benz Söhne with his sons Eugen and Richard. The mainstream concern he left behind immediately developed competition ambitions to publicize its wares.

In 1907 Benz (the original concern) built cars with dual-ignition inlet-over-exhaust engines of 8- and 8.9-litres capacity; this 60 hp model became a favourite of Prince Henry of Prussia himself and brought works driver Fritz Erle the important Herkomer Trophy – the Herkomer was in effect one of the earliest sports-car competitions.

Victor Hémery was employed as another factory driver, and he was second in one of these 60s in the Coppa Florio race, ahead of team-mate René Hanriot's sister car. By 1908 Benz were heavily committed to motor racing, and their four-cylinder 12½-litre cars, designed largely by Frenchman Louis de Groulart, served them very well. Hémery won the incredibly rough St Petersburg-to-Moscow race on the model's debut, covering the 438 miles at a staggering average of 51 mph. In the Grand Prix at Dieppe, Benz were robbed of victory only by tyre failures and by an eye injury to Hémery who finished second behind Lautenschlager's victorious Mercédès (Mercédès and Benz did not merge until 1926). Hanriot's Benz was third in the Grand Prix with Erle seventh. The cars later went to America where they were raced with enlarged engines, Hémery being timed at over 100 mph in the Grand Prix race at Savannah and finishing narrowly second behind a Fiat. Hanriot was third yet again. In 1909 David Bruce-Brown covered the mile at 109 mph in his Benz, now with 15 litres beneath the bonnet.

Lion-Peugeot Voiturette

1908, France

This odd-looking little car was another of the 'Formula 2' voiturettes which abounded in the years before World War I. It was built by Robert Peugeot's company, founded as an offshoot of the main Peugeot family concern still active today.

The single-cylinder 'thumpers' made their major competition debut in the 1906 Coupe de l'Auto, finishing third, and were third again in the same event in 1907.

The way to increase power was in those days believed to be simple enlargement of the engine. The cylinder bore was considered to be quite close to practical and Formula limits, and so the stroke was lengthened enormously, producing the immensely tall engine which characterized the Lion-Peugeots and so many similar cars of this period. To compensate for the height of an engine with a 100 mm bore allied to a 170 mm stroke, the Lion-Peugeots were low-slung in chassis and body form.

Finally, in 1909, Lion-Peugeot developed the model which became legendary – fitted either with a 100 mm × 250 mm single-cylinder or an 80 mm × 192 mm narrow-angle V-twin. The singles had bags of low-speed torque available to punch the cars very rapidly away from slow corners, and they proved admirably suited to most racing circuits of the time. They were driven by Jules Goux to beat the Hispanos into second place in the Catalan Cup, and he won the Sicilian Cup as well. Team driver Guippone beat Goux in Lion-Peugeot's 1–2 success in the Coupe de *l'Auto*, and so

By 1910 long-strokery reached its height in the twin-cylinder Lion-Peugeot voiturettes, as shown here. The driver and riding mechanic had to look round, rather than over, their car's bonnet.

they went on. Goux also drove the formidable Type VX5 model, fitted with an 80 mm × 280 mm engine, which could achieve 95 mph and whose driver was forced to look around rather than over the phenomenally tall power unit. Michaux also produced a 3440 cc V4 engine of 65 mm × 260 mm and more successes came Lion-Peugeot's way.

By 1911, however, the two breeds of Peugeot were reunited and from that date competition Peugeots had in-line engines, conventional looks and great mechanical sophistication.

'Coupe de *l'Auto*' Hispano-Suiza *1910, Spain*

While the age of monsters lingered on in heavy-car racing, and the world trade recession caused the abandonment of the Grand Prix race from 1909 to 1911, technical innovation was fostered by the Coupe de l'Auto race for voiturettes.

We have already seen how Sizaire-Naudin and Lion-Peugeot fared, but the Spanish-based, French-prepared and Swiss-engineered Hispano-Suiza was perhaps more significant than either in the technical sense. Swiss engineer Marc Birkigt had produced the La Cuadra car in Barcelona in 1900 and was responsible for the Castro car made in that city from 1901 to 1904. The first Hispano-Suiza of 1904 was actually the last Castro renamed. Birkigt would accept second best in nothing, and his Hispanos became rapidly established as supremely well-engineered and sophisticated cars.

It was in 1909 that the company made its first serious competition effort with a team of 65 mm × 140 mm 1852 cc four-cylinder cars which ran in the Catalan Cup. They proved unsuccessful, but in the following Coupe de *l' Auto* they were placed 5–6–7 and then went on to major hillclimb class victories.

When Lion-Peugeot won the 1910 Catalan Cup, Paul Zuccarelli's now 170 mm-stroke Hispano-Suiza finished third. For the Coupe de *l'Auto* Birkigt lengthened the cars' piston stroke to 200 mm and with 60 smooth horsepower, a great spread of power throughout the engine's rev range and a well-developed chassis, the Hispano-Suiza was victorious. Paul Zuccarelli won, Chassagne was second and Pilliverdier sixth.

In 1911 Marc Birkigt developed a new overhead-camshaft engine which departing engineers Zuccarelli and Ernest Henry subsequently took to Peugeot, and for the 1912 Coupe de *l'Auto* Hispano's Swiss genius planned a team of supercharged cars – ten years before supercharging was to make its mark in motor racing.

Bugatti Type 13

1911, France

Ettore Bugatti is now a legendary figure in motoring and motor-racing history. The Italian-born French-domiciled artist-engineer produced some of the most aesthetically appealing classic cars of a classic era, and Bugatti as a marque was for many years top of the most-race-wins league.

Bugatti's reputation in competition was founded largely by his Type 13 model, which became known subsequently with its derivative the Type 22 as the 'Brescia', after the Italian venue at which it scored its first major multiple victory.

Bugatti himself had achieved some sporting laurels with cars of his own design but built under other names before he established his own independent factory at Molsheim, Alsace, in 1910. The following year, 1911, saw what became known as the *Grand Prix des Vieux Tacots* ('the Old Crocks' Grand Prix') run on a circuit at Le Mans. Since major manufacturers were not ready for a resumption of formal Grand Prix racing, many dealers and interested sportsmen ran what cars they could find in this event. There, in amongst the mighty chain-drive Fiats, was a team of spidery little Bugatti Type 13s, and Ernest Friderich drove one home into second place overall.

When racing resumed after World War I, modified 16-valve $1\frac{1}{2}$-litre Type 13s reappeared under Bugatti factory colours, Friderich winning the Voiturette Grand Prix at Le Mans in 1920. This brought much publicity to the new marque, and at Brescia in 1921 Friderich led a Type 13 1–2–3–4 grand-slam victory which won the type its immortal name.

When compared to the other crude voiturettes of the pre-1914 period – the Hispano-Suiza apart – the Bugatti Type 13 was truly a luxury car in miniature, and although it offered only 25 bhp or so, this power was produced in a manner which anticipated the smooth and lively high-efficiency light cars of the thirties. No wonder the Bugatti made such a tremendous impact in 1911, and was good enough to continue very effectively far into the twenties in modernized form.

'Coupe de *l'Auto*' Delage

1911, France

Louis Delage was a flamboyant and wildly extrovert character, and he came to prominence in racing at virtually the instant he commenced motor manufacture – one of his voiturettes with a single-cylinder De Dion engine coming second in the 1906 Coupe de *l'Auto* race.

By 1908 Albert Guyot could be provided with a twin-cylinder De Dion-engined Delage with which to win the Coupe. Delage should have gained much credit for this victory, but since he badly needed De Dion's bonus money he insisted full credit be given them. Work proceeded on producing Delage's own power unit and by 1911 when the factory again contested the Coupe de l'Auto they were using their own three-litre four-cylinder horizontal-valve engines allied to unusual

five-speed gearboxes with an over-drive top ratio.

Paul Bablot, René Thomas, Albert Guyot and Louis Rigal were hired to drive the cars and the first three finished 1–3–4. Bablot averaged no less than 55.3 mph in the winning car, a noble performance which proved to the buying public that the Delage factory could produce cars which were both fast and reliable.

Not that they were very pretty, for the high-set cranked exhaust pipe and ugly bolster fuel-tank on the tail did nothing to improve the car's aesthetic qualities. In deference to regulations demanding bore and stroke dimensions no more extreme than 78.1 mm × 156.2 mm for the Coupe, Delage had chosen 80 mm × 149 mm and this posed practical problems in filling the cylinder rapidly with gas on each induction stroke. Hence the unusual valve and exhaust layout. The exhaust emerged from the top of the cylinders due to Delage's chosen use of valves on each side of the cylinder head. The engine gave about 50 bhp at 3000 rpm which was not outstanding, but the cars were very well prepared and driven with great skill, and the five-speed gearboxes allowed the driver to select the most advantageous part of the engine's rev band at each point on the testing circuit.

The magnificent Coupe de l'Auto-*winning Delage of 1911 survives to this day in the collection of Sir John Briscoe.*

Grand Prix Fiat

1911, Italy

Of all the early Grand Prix cars perhaps the most sensational-looking in terms of sheer brutishness is the Italian Fiat, built by the Fabbrica Italiana dell'Automobili Torino. In 1907 the Fiat works driver Felice Nazzaro had won the Grand Prix in a 16,286 cc 120 hp chain-drive model. In 1908 the S61 Grand Prix model was a more modest 10,087 cc variant. In 1911 the S74 appeared, with 14,137 cc from a four-cylinder 150 mm × 200 mm engine, chain-driven of course, and fitted with the most sketchy of racing bodies – merely a hutch-like box covering the engine up front, two tiny seats for driver and riding mechanic, and behind that the transverse cylinder of the fuel tank and a rack for four spare tyres.

It was with one of these cars that Louis Wagner fought long and hard against Boillot's new-vogue Peugeot in the 1912 Grand Prix at Dieppe; but in 1911 the S74 had already won the American Grand Prize race for New Yorker David Bruce-Brown at Savannah, Georgia. Bruce-Brown was invited to drive the car again in the Grand Prix at Dieppe and he led the first day's racing, only to be disqualified on day two for infringing the regulations by refuelling the car away from its allocated service pit.

Bruce-Brown was subsequently killed during practice for the 1912 American Grand Prize race at Milwaukee, Wisconsin, when a tyre burst and threw his big S74 out of control during some high-speed practice laps. Bruce-Brown was only just 22, and had been a national celebrity after winning the Grand Prizes of 1910 in a Benz and 1911 in the Fiat. Now he was a much-mourned national hero.

The mighty chain-drive Fiats continued to be raced in America, notably by 'Terrible Teddy' Tetzlaff, and by Caleb Bragg (normally driving an S61). At least one S74 survives to this day.

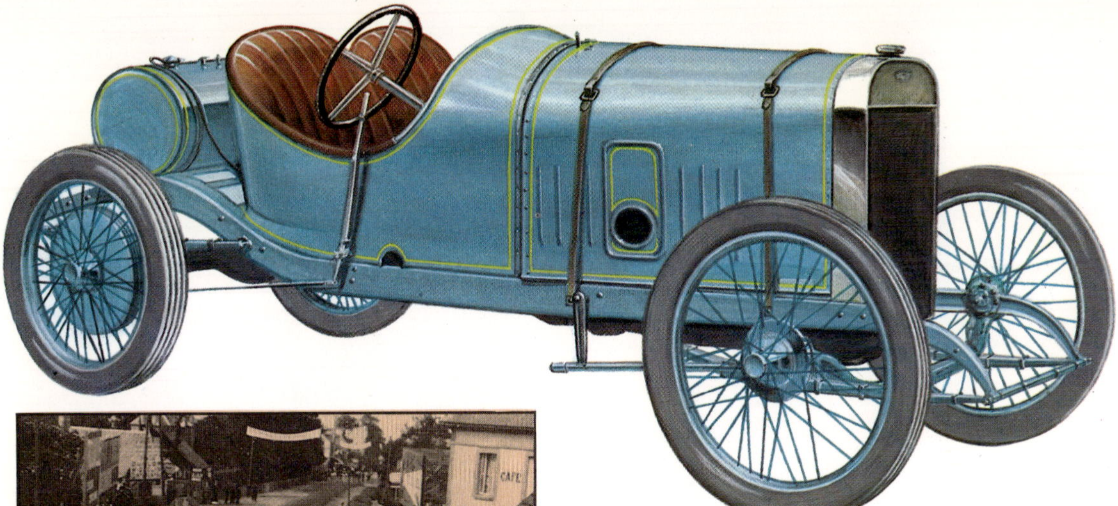

◄*Georges Boillot and his riding mechanic urge their phenomenally successful twin-overhead-camshaft, four-valve-per-cylinder Peugeot on to victory at Dieppe in the 1912 Grand Prix de l'ACF. The long circuit was typical of the time, passing through many villages where the pavements were boarded off, as seen here.*

Grand Prix Peugeot

1912, France

Today all self-respecting racing engines feature twin overhead camshafts actuating four valves in the head of each cylinder – two inlets allowing the fuel/air mixture to enter, and two exhausts allowing the burned products to escape. In 1912 Peugeot introduced this basic layout to Grand Prix racing with phenomenal success, their complete dominance lasting into 1913, challenged by Mercédès only at the last gasp in 1914.

Paul Zuccarelli and Ernest Henry had come to Peugeot with Birkigt's Hispano-Suiza ideas, and under the direction of the famous 'Charlatans' team-driver-engineers Zuccarelli, Georges Boillot and Jules Goux – Henry produced the design which was the true progenitor of the modern racing car. The Henry Peugeots featured shaft-drive instead of chains, and 16-valve four-cylinder engines with twin overhead camshafts driven by bevels. One model for the Coupe de *l'Auto* had bore and stroke of 78 mm and 156 mm respectively, to produce three litres, while the corresponding Grand Prix car had 7.6 litres and an output of around 130 bhp at 2200 rpm.

In the Grand Prix at Dieppe the Coupe de *l'Auto* race was run concurrently, the small-capacity cars mixed in with the Grand Prix cars. The 7.6 Peugeot was a kind of halfway-house between the little voiturettes and the 12–15-litre Grand Prix norm with overweight chassis and normally chain drive. The result of the Grand Prix was a triumph for new over old as Boillot drove his 7.6 brilliantly to beat Louis Wagner's *tacot* Fiat with its 14-litre engine and chain-drive rear end.

The three-litre Peugeot voiturette was retired by René Thomas in this event but Zuccarelli later won the Grand Prix de France at Le Mans with it, and Goux the Coupe de la Sarthe in a 7.6. In 1913 Goux won the money-rich Indianapolis 500-Miles in America with one of these Peugeots, and in 1913 a new 5.6-litre Grand Prix car carried Boillot to his second successive Grand Prix victory. He was the first man to win two Grands Prix, and became a French national hero. For 1914 regulations demanded only 4.5-litre engines, and with four-wheel brakes (unusual at that time) Boillot fought heroically to achieve his hat-trick until the car failed.

'Coupe de *l'Auto*' Sunbeam

1912, Great Britain

While French marques generally dominated the early years of motor-racing, with occasional intrusion from Germany and Italy, Britain came nowhere in the international league table. But in 1907 the Brooklands Motor Course was opened near Weybridge in Surrey, and with its impetus competition-orientated companies began to make great strides forward, particularly in the design of aerodynamic bodywork demanded by the flat-out high-speed character of the Brooklands track.

Sunbeam began competition in that same year and under the direction of the Breton engineer Louis Coatalen their cars progressed from strength to strength. In 1911 Coatalen produced a Coupe de *l'Auto* car to take on the French and Spaniards at their own game. It was a three-litre car with monobloc (all four cylinders cast into the one piece of metal) engine derived from the production 12/16hp model. It retired from the prestigious French race, but Coatalen had taken a good hard look at the opposition, and returned in 1912 with a team of four much improved machines.

The 1912 cars still retained side-valves, unlike the overhead valves in vogue with most of the Europeans, but their neat little three-litre engines delivered a reliable 74bhp. They had Brooklands-derived long-tailed streamlined bodyshells to gain a little aerodynamic advantage over their slab-tailed foreign opposition, and their top speed was up around 90mph. If only they could survive the race distance, they were bound to be very competitive. In the event, as the Coupe de l'Auto was run concurrently with the Grand Prix they showed very well not only against their three-litre rivals but also against the heavy cars of the Grand Prix itself.

Rigal, Resta and Caillois actually finished 1–2–3 in the Coupe de l'Auto to achieve a British victory which stunned the French establishment, and their cars actually finished 3–4–5 in the Grand Prix! The winner averaged 65.35mph in the Coupe and this same model later collected over 30 class and outright world speed records, achieving 99.45mph for the mile. It was the start of great things for Sunbeam – a now devalued name.

One of the fabulously successful 1912 Coupe de l'Auto *Sunbeams as it survives today, at the British National Motor Museum, Beaulieu.*

Dario Restá hurtling round Dieppe's typically tree-lined country roads in the 1912 Coupe de l'Auto *Sunbeam.*

The Coupe de l'Auto *Sunbeam's four-cylinder, three-litre engine was modest, neat and powerful.*

'Indianapolis' Mercer
1913, United States

The Mercer Automobile Company of Trenton, New Jersey, USA, was founded in 1910 and began competition upon the introduction of its rakish Raceabout model in 1911. The marque's early successes were achieved by stripped production chassis but from 1913 a special competition version–the Type 45–was fielded for contracted works drivers.

In 1912 'Hughie' Hughes had finished third at Indianapolis in the annual 500-Miles Sweepstake event–which was at that time and has continued to be the world's richest individual motor race. His Mercer's engine was smallest in the race, and the car averaged 76.3 mph for the 500-Mile distance. Many other victories came Mercer's way on road circuits like Santa Monica, California, and Elgin, Illinois, and in 1913 the 7292 cc 150 bhp Type 45 racer joined the Type 35-engined 300 cubic-inch class (4916 cc) Model F competition car in the front-line.

Spencer Wishart in a Type F was second at Indianapolis behind Goux's Peugeot, while Ralph de Palma–possibly the most successful racing driver of all time in terms of sheer race victories–won at Elgin, also in a Type F, and achieved three wins from five races entered at San Antonio.

In 1914 Eddie Pullen won the American Grand Prize race at Santa Monica in a Type 45–the first time that an American car had won this event. Naturally this paid enormous promotional and publicity dividends to the Mercer company, and in celebration the whole Mercer works was actually given a half-day holiday–an unheard-of gesture at that time.

Mercer continued racing until America entered World War I in 1917, and no further competition programmes were attempted upon the return of peace. Still, Mercer are worthy representatives of the budding age of the American competition car, and America's own brand of enclosed artificial track racing was to breed a type of extremely powerful, very fast, and immensely reliable racing machine which in many ways was ahead of European practice during the twenties. Unfortunately the virtual collapse of open-road racing in the USA in that same period meant that the nation could not mount a Grand Prix challenge.

Grand Prix Mercédès
1914, Germany

When the Automobile Club de France issued its regulations for the Grand Prix of 1914 it opted for the first time to limit outright engine capacity, whereas earlier limits had been applied to overall weight and fuel consumption. The limit was to be 4.5 litres, and the result was perhaps the greatest Grand Prix race of all time: 14 manufacturers made special cars for this one event. Prominent among them were Peugeot in France and Mercédès in Germany. Mercédès had won the 1908 Grand Prix upon dogged reliability and good organization, while in 1912 and 1913 Peugeot, with the immense driving talent of the flamboyant Georges Boillot, had won the race on flair and technical innovation and were now determined to perform the hat-trick *pour la France* . . .

In Stuttgart, the Mercédès engineers worked on a new 4.5-litre shaft-drive design. The engines derived from a 1913 six-cylinder unit with overhead valves and camshaft, hemispherical combustion chambers (this shape promoted very efficient burning of the fuel/air charge), dual magneto ignition and four spark plugs per cylinder to enhance combustion efficiency still further. They adopted a method of engine construction from contemporary Mercédès aero-engines, building-up the unit from steel forgings, with the ports welded-up on bare cylinder heads and sheet water-jackets then welded carefully round the whole assembly. This technique replaced the heavyweight casting used previously and eliminated any chance of failure through the porosity or weak spots often found in castings.

While these new engines developed around 110 bhp and gave a top speed of some 100 mph evidence of Mercédès' forethought and atten-

The four-cylinder engine of the 1914 Grand Prix Mercedes was copied in essential details by Rolls-Royce for their World War I aero-engine designs.

One of the $4\frac{1}{2}$-litre 1914 works Mercédès wheeling round the Lyon-Givors circuit's Piège de la Mort *corner during that great Grand Prix.*

tion to detail was also found in the transmission line, where each axle shaft in the rear axle assembly was set at 1° 20′ from the horizontal; this gave an inward camber on the rear wheels and so improved their grip in cornering, the inclination on the shafts demanding individual crown-wheels and pinions to drive each one. What was more, each pinion gear was machined integrally with the halfshaft and each car took with it to France enough half-shafts to give a choice of six final-drive ratios – from 2.2 to 2.7:1. Each was to be tested around the Lyon-Givors circuit, and that which produced the best lap time would be fitted for the race. This spare-no-pains, spare-no-expense attitude to motor-racing success was to become typical of Mercédès when the company returned to racing in earnest in the twenties; the method of engine construction in fact survived into their Grand Prix and sports cars of the mid-fifties.

At Lyon, Max Sailer set the initial pace, tearing round ahead of Boillot's Peugeot before blowing up. Lautenschlager and Wagner then strode implacably forward, to harry Boillot until the Peugeot failed on its final lap. Lautenschlager finished first at an average of 65.38 mph from Wagner and Salzer in sister cars, their German 1–2–3 triumph being received in stony silence by the partisan French crowd. Three weeks later France and Germany were at war – for reasons unconnected with motor-racing and commercial prestige.

The Grand Prix Mercédès were saved during the war; one was studied by Sir Henry Royce in Britain, another was raced extensively by Ralph de Palma in the USA and passed from him into the Packard experimental department. The Rolls-Royce aero-engines which powered the DH4, DH9 and Bristol fighters during that war owed much to the Mercédès.

Christian Lautenschlager's actual winning car survives in England to this day, and is still raced on occasion.

Great Mercs together – on the right the actual winning car from the 1914 Grand Prix and on the left a rebuilt 1937 W125, the most powerful Grand Prix car ever raced. Both survive today owned by British enthusiasts.

Land Speed Record Delage Type DH V12

1923, France

Louis Delage was always seeking potential publicity and promotional exercises; in addition to spending a very considerable fortune on the racing cars which promoted the image of his high-quality tourers, he also invested in some specialized sprint cars for high-speed straight-line work.

Immediately after World War I he authorized construction of a six-cylinder 5136 cc sprint and hillclimb car in which René Thomas won the Mont Ventoux climb in 1922 and 1923, breaking the mountain record both times. At La Turbie in 1923–24 he repeated these successes and shattered the record at the Zbreslav-Jiloviste climb in Czechoslovakia to spread Delage fame abroad. A second car was also built with a six-litre twin-overhead-camshaft engine and this was similarly successful, making fastest times at both Mont Ventoux and La Turbie.

Meanwhile, in 1923, Delage produced a massive V12-cylinder car in which two banks of six cylinders each were coupled in 'V' formation upon a common crankcase. This mighty assemblage displaced 10,688 cc and produced around 300 bhp.

René Thomas drove the car in some major sprint events, such as the hillclimb run every year at the time of the Paris Motor Show at Gaillon, a village between Paris and Rouen. At Arpajon in 1924 Thomas held the car bravely to record 143.29 mph in a sprint meeting, which figure qualified the car as holder of the World Land Speed Record. Within minutes Ernest Eldridge's mighty chain-drive Fiat special *Mephistopheles* (of over 20 litres capacity) went faster but Thomas protested the car since it lacked the regulation reverse gear. This was not very sporting but was typical of French motoring attitudes at that time. Eldridge laboured for two days to provide his special with a reverse gear, and in another run at Arpajon demolished the big Delage's figures.

In 1928–29 Captain Alastair Miller brought the two smaller Delage sprint cars to Brooklands and Thomson and Taylor, the Brooklands tuning and preparation specialists, brought over the big V12. It survives in England today.

Fiat Tipo '805-405 Corsa'

1923, Italy

In 1922 a new capacity limit of two litres was applied to Grand Prix cars. The following year, Fiat emerged with the first successful

supercharged Grand Prix car. Instead of relying upon atmospheric pressure and the engine's own suction to inspire fuel/air mixture, the Italian car used a compressor pump, or supercharger, to force fuel/air into the engine. The more mixture that could be forced into the cylinders, the more could be burned and therefore the more energy extracted from each individual charge.

Chadwick in America and Hispano-Suiza in Spain had toyed with supercharging before the Great War, and in 1923 Fiat made it fashionable. The straight-eight engines had bore and stroke dimensions of 60 mm and 87.5 mm respectively to give 1979 cc, and with the Wittig vane-type supercharger they developed a quoted 130 bhp at 5500 rpm. With a Roots supercharger 150 bhp was claimed. On their first outing at Strasbourg for the 1923 French Grand Prix the team of 805s all retired, apparently due to road grit and dust drawn into the Wittig superchargers, scoring their internal surfaces and creating loss first of supercharge pressure and then of lubricating oil, and finally internal break-up of the engines.

Fiat put the cars right in time for their national Grand Prix at the new Monza Autodrome just outside Milan, on which artificial circuit the now Roots-supercharged 805s carried Carlo Salamano to victory before an hysterically enthusiastic crowd.

In 1924 Fiat continued to race these cars, Pietro Bordino finishing fourth in the Targa Florio, but all three cars retired from the French Grand Prix and entries were scratched from the Italian. At this time several of Fiat's top engineers had been lured away to join the rival Alfa Romeo company, notably via introductions by a young man named Enzo Ferrari.

The Fiat board was justifiably incensed by the team's being used as a technical training school to foster other companies' racing development. When a 'Fiat-copy' Sunbeam won the French Grand Prix of 1923, and the supercharged '805-copy' Sunbeam won at San Sebastian in 1924 – Alfa Romeo having won the French Grand Prix that year – Fiat withdrew from racing.

Brooklands Lanchester

1923, Great Britain

This was one of the most unusual of the many specials devised between the wars for the high-speed banked Brooklands Outer Circuit. Many of them, such as Count Louis Zborowski's 'Chitty Bang Bangs', and Cobb's Napier-Railton (see page 50) were aero-engine powered. Tommy Hann's Lanchester coupé was rather different. The car upon which he based his striking Brooklands special was a 1911 25 hp Landaulet, running originally as an aerodynamic tandem two-seat saloon and finally as an open single-seater. Hann named the car 'Softly-Catch-Monkey' in its open form, although the name was not a patch on the original saloon version's imaginative title, 'Hoieh Wayaryeh Gointoo' – apt commentary upon the driver vision afforded by the coupé's arrow-slit windows.

The engine was a 102 mm × 102 mm, 3334 cc four-cylinder unit with horizontal valves, and the car used the Lanchester system of cantilevered suspension, pressure-lubricated epicyclic semi-automatic gearbox and bevel-drive back axle. Tommy Hann found the car 'deafening' to drive, hence the subsequent conversion into an open racer.

Grand Prix Sunbeam

1923, Great Britain

Louis Coatalen of Sunbeam is alleged to have held the pragmatic view that ''Ee ees a wise man who copy wizout altair'. In the twenties, if Coatalen saw an outstanding racing car produced by another company he would unashamedly copy it, paint it green and call it a Sunbeam, or perhaps blue and call it a Talbot, by virtue of the complex Sunbeam-Talbot-Darracq Anglo-French grouping of that period.

Coatalen had long admired the work of Ernest Henry and in 1922 he hired him to produce a new design for Sunbeam. Unfortunately Henry's great creative days – or perhaps more realistically his days as a gifted draughtsman capable of translating others' ideas efficiently into the metal – were over.

With Peugeot, Henry had led the way in 1912. With Ballot he had produced the first European Grand Prix straight-eight in 1919. Now at Sunbeam he chose a twin-overhead-camshaft 16-valve four-cylinder layout, but small piston area and a rev limit of only 4400 rpm rendered the cars ineffective, and they retired from the Grand Prix with valve failures. This Sunbeam model was Henry's last complete design, and with the 1923 Mercédès the last four-cylinder Grand Prix car to appear for a quarter-century.

With his faith in Henry shattered, Coatalen looked to the engineers who had brought Fiat their French Grand Prix victory at Strasbourg in 1922, when the veteran Felice Nazzaro's Tipo 804 had won using a six-cylinder, 65 mm × 100 mm, 1991 cc engine. Coatalen succeeded in attracting engineers Vincenzo Bertarione and Walter Becchia from Fiat, and Bertarione produced the 1923 Grand Prix Sunbeam as a six-cylinder of 67 mm × 94 mm, achieving 102 bhp at 5000 rpm compared to the 1922 Fiat's 112 bhp at 5000 rpm.

At Tours for the French Grand Prix of 1923 the new supercharged Fiats failed and Henry O'Neill de Hane Segrave won for Sunbeam in his 'Fiat in Green Paint'. It was to be the last important GP victory for an Englishman in an English car until the Syracuse Grand Prix of 1955 fell to Tony Brooks and his Connaught.

At the Touraine pits during that historic Tours Grand Prix of 1923; Sunbeam refuels.

Riding mechanic Paul Dutoit has ducked down out of sight as Major Segrave takes the works Sunbeam to victory at Tours, in 1923.

Grand Prix Voisin

1923, France

In many ways the season of 1923 was extraordinarily innovative: Fiat adopted supercharging in their Grand Prix cars, Benz produced the incredible *Tropfenwagen*, and in France Gabriel Voisin – the former pioneer aviator and constructor of some 10,000 aircraft during the war years – promoted his motor-manufacturing business by producing the first 'monocoque' Grand Prix car.

Today every Grand Prix car uses a monocoque chassis, this term being mongrelized from Latin and Greek to mean 'single-shell'. Whereas chassis frames had traditionally consisted of a ladder-like framework of wood or steel upon which engine, gearbox, axles and body were mounted, in a monocoque the bodywork is formed in stressed sheet panelling, and of itself is sufficiently strong and rigid to carry engine, gearbox and suspension loads. A rolled and glued sheet of paper is a monocoque structure. An engineer named Louis Béchereau had devised this type of structure, made from laminated and glued wooden strip, in his Deperdussin racing aeroplane in 1912.

Eleven years later, Gabriel Voisin applied the idea to his Grand Prix cars for the French event at Tours. His two-litre six-cylinder sleeve-valve engines were underpowered, offering 75 bhp at best, and he sought to compensate by saving weight and wind-resistance. So he formed his cars around a plywood and sheet-metal monocoque chassis structure which doubled as the bodywork in places. In style the cars resembled an about-faced aerofoil in profile, narrow edge forward and deepest point just aft of the two-man cockpit. The undertray was entirely flat, but unfortunately an ungainly box had to be raised in the front to enclose the engine. Sadly for Voisin his sophisticated cars still proved slow and – worse – unreliable, and the best his team could salvage from the Grand Prix was fifth place.

Not until 1962 would the monocoque structure make an impact upon Grand Prix racing. Voision paid the price of being a prophet before his time.

Grand Prix Sunbeam

1924, Great Britain

His 'Fiats in Green Paint' having won at Tours in 1923, for 1924 Louis Coatalen and his Sunbeam company tackled Grand Prix racing with a supercharged version of their winner.

The engines were still basically Bertarione's 67 mm × 94 mm 'sixes' mounted in modified chassis with the wheelbase extended to 8 ft 6 in and the tracks at front and rear proportionately wider. A disc clutch and four-speed transmission were adopted while in the Wolverhampton test houses, Coatalen and his technical director Captain Irving experimented extensively with a Roots vane-type supercharger blowing compressed air through the Solex carburettor. Eventually they decided that the system was more effective with the supercharger drawing fuel/air mixture from the carburettor, and so Sunbeam became the first European marque to compress mixture rather than air alone. This subsequently became standard practice; in this instance Sunbeam were not mindlessly copying, but actually improving on what had gone before.

The Sunbeam in this form peaked at around 138 bhp at 5500 rpm, and in general characteristics the engine proved slightly stronger than that of the P2 Alfa Romeo at Lyon, where the British cars took the lap record; but the Sunbeams were dogged by magneto trouble and lost the race to the Italians.

Segrave compensated by winning the San Sebastian Grand Prix in Spain later in the year, the last Grand Prix victory of any significance by an all-British car/driver combination until Tony Brooks's Connaught won at Syracuse in 1955; Segrave's Tours victory of 1923 was to be the last major all-British Grand Prix win until 1957.

A team of British green – the Sunbeams at Lyon-Givors, 1924.

Protagonists at Lyon, 1924, with Segrave's Sunbeam (left) and Campari's Alfa Romeo (right) rounding the tricky Piège de la Mort *corner above the pits.*

Alfa Romeo P2

1925, Italy

The Societa Anonima Lombarda Fabricca Automobili of Milan, Italy, initially made its name in motor racing with cars designed by Ing. Giuseppe Merosi. In 1915 ALFA became part of industrialist Nicola Romeo's empire, and the cars became known then as Alfa Romeos.

One of the company's racing drivers in the early twenties was Enzo Ferrari and this thrusting and aggressive young man was used as an entrepreneur in that period to attract top-line racing car design staff to Alfa from Fiat of Turin. In this, Ferrari was extremely successful, his prize bag being Ing. Vittorio Jano. At Alfa Romeo Jano studied the existing Merosi-designed, production-derived competition cars, and after the failure of an exploratory P1 Grand Prix model he devised the outstanding P2 for 1924.

Jano used Merosi's abandoned 1923 six-cylinder Grand Prix design as a test bed for super-charging experiments, and from it developed the 61 mm × 85 mm supercharged two-litre unit for the car which they called the P2, a straight-eight with twin overhead cam-shafts. The car won the very first Grande Epreuve it contested, at Lyon in 1924 – Giuseppe Campari won the race and an outsize Lyon sausage for his trouble. Campari was a burly, opera-singing driver of outstanding merit, and his team-mate Antonio Ascari was, if anything, an even greater racing driver. That October, Ascari won the Italian Grand Prix at Monza in a P2 producing a claimed 145 bhp at 5500 rpm and in 1925 power output achieved 155 bhp. Ascari won the Belgian Grand Prix at Spa-Francorchamps, but at Montlhéry he ran out of road and was killed while leading the French Grand Prix – the whole team was withdrawn. Brilli-Peri won the Italian Grand Prix that year in another P2 to clinch the World Championship title for Alfa Romeo. Alfa's badge has worn these laurel wreaths ever since.

The six P2s continued to achieve great success in minor events long after the $1\frac{1}{2}$-litre Grand Prix Formula was applied in 1926, bringing Achille Varzi to prominence as a great driver as late as 1930.

Delage 2LCV V12

1925, France

While Fiat were being increasingly frustrated in Grand Prix racing, Alfa Romeo and Sunbeam were poaching talent and innovation, and Voisin and Benz were tackling new methods with little good fortune, Louis Delage's company grew in stature and confidence until the mid-twenties saw it producing some of the most complex, most expensive and most brilliant Grand Prix cars the sport has ever seen.

Delage himself had been grievously disappointed by the failure of his 1914 Grand Prix car, which had been intended to take the marque as triumphantly into this class as others had into the Coupe de *l'Auto* Voiturette competition. The 1914 car had featured four-wheel brakes, four carburettors for the four-cylinder engine, five-speed gearbox and desmodromic – mechanically closed – valves. Now the patron's cousin, Ing. Charles Planchon, designed and built the second Delage Grand Prix contender, nine years after the first.

The two-litre V12 Delage 2LCV was immensely complex, yet built to a frantic time schedule. Drawings were not completed until March 1, 1923, leaving 120 days to build the team cars in their entirety, and to test and develop them. The engine was built and tested in less than eight weeks, the tiny cylinders each measuring 51.4mm bore × 80mm stroke and connected to a seven-roller-bearing crankshaft. There were twin overhead camshafts on each bank of six cylinders, and in original form these V12 engines gave about 120bhp. They resided in orthodox chassis with servo brakes, were very light and compact but too new to do well at Tours.

For 1924 Albert Lory developed the V12, experimentally trying twin superchargers, but at Lyon the team cars ran with atmospheric induction. They were placed 2–3–6 when the Sunbeams and Alfa P2s suffered problems. In 1925 the superchargers were standardized, and with 190bhp the Delages came 1–2–3 in the Spanish Grand Prix in the absence of Alfa competition and Divo/Benoist shared the winning V12 in the French Grand Prix at Montlhéry, following Alfa's retirement. Louis Wagner was second, and Delage were the toast of France.

◄*Louis Delage with his V12-engined Grand Prix car at the Targa Florio of 1926.*

►*1926 Bugatti Type 35 in a VSCC club race at Silverstone.*

Bugatti Type 35

1926, Italy

Ettore Bugatti believed throughout his life that the cars he sent from Molsheim to race on the circuits of the world should be similar to those he offered for sale to the general public. Fiat, Delage and Alfa Romeo all considered such restraints impracticable and unrealistic. Bugatti was to prove them wrong to a considerable extent.

For the Grand Prix of the Automobile Club de France at Lyon in 1924, he unveiled his Type 35, perhaps the greatest classic beauty of its era. He considered supercharging an Italian trick and an infringement of the two-litre capacity regulations, and since he was Italian by birth few could criticize his comments on jingoistic grounds.

His 1924 Type 35 used a vertical-valve straight-eight engine of 60 mm × 88 mm, 1900 cc, with a built-up eight-section crankshaft revolving in five main bearings. In both construction and effect the Bugatti chassis was a revelation, its side members being tapered fore and aft to offer the greatest section and therefore greatest strength at the points of maximum stress, while the engine contributed materially to its rigidity. Semi-elliptic front leaf springs passed through forged slots in the front axle, which was itself a masterpiece of craftsmanship with its bowed centre section and tapered, upswept extremities. Bugattis had been typified by their quarter-elliptic rear springs reversed forward onto the axle, but in the Type 35 they curved inwards at the tail and were concealed within the beautifully proportioned bodyshell, with its Bugatti horseshoe radiator.

The wheels defied 15 years' acceptance of the spoked-wire type, being of cast alloy with spokes integral with both the rims and the brake drums, thus rendering brake shoes easily replaceable at tyre-change stops. The Type 35 displayed superb handling, magnificent torque and braking, and an essential charm. All these qualities went to make up a revered motor-car.

In 1925 the Type 35 made its real mark winning the first of five successive Targa Florios, and in 1.5-litre form the Italian Voiturette Grand Prix. In 1926 Bugatti supercharged it and won the European Championship, and from 1928 to 1930 the Bugatti 35B in supercharged 2.3-litre form had things more or less its own way in racing everywhere.

The Bugatti cast-alloy wheel, integral with rim and brake drum.

Delage Straight 8

1926, France

After the twelve-cylinder complexities of the 1923–25 two-litre Grand Prix car, Louis Delage authorized Albert Lory to produce a new design for the new 1.5-litre Grand Prix Formula in force during 1926–27. Lory chose to produce another engine of watch-like complexity, with eight cylinders in line and twin overhead camshafts driven from a multi-wheeled gear train at the front of the unit.

While Lory's supercharging of the Planchon-designed V12 cars had employed superchargers mounted on each side of the crankcase, he laid out his 1926 straight-eights with two superchargers slung on the left-side of the crankcase and operated from the cam-drive gears up front. This move placed the exhaust system on the right side of the car with number eight exhaust port discharging very close indeed to the driver's feet.

Since many fatal accidents had involved the riding mechanics mandatory in Grand Prix cars up to 1925, the regulations were changed in 1926, demanding two-seater-width bodies still but allowing the driver to race single-handed. Lory consequently sat the driver of his new car down beside the angled propellor-shaft instead of on top of it, the shaft passing through what would have been the mechanic's lap, and so lowered the new car without infringing the new 31.5in minimum-width rule. This in turn reduced the car's frontal area, reducing the wall of air it had to force before it, and so contributing materially to its aerodynamic efficiency.

Mechanically the 1926 cars proved extremely efficient, the new engines delivering around 160bhp at 6500rpm. In their first season they burned their driver's feet unbearably, but Edmond Bourlier/Robert Seneschal were second at San Sebastian, and Wagner/Seneschal won at Brooklands in the first RAC Grand Prix.

For 1927 Lory revised the cars to place the exhaust on the opposite side to the driver, and they dominated the Grand Prix season, Robert Benoist winning everything in sight. Louis Delage retired his team from racing in 1928 but as late as 1936 Dick Seaman came to prominence in one of these fabulous cars.

The great Dick Seaman winning the JCC 200-Miles at Donington Park in 1936 in his rebuilt nine-year-old Delage straight-eight.

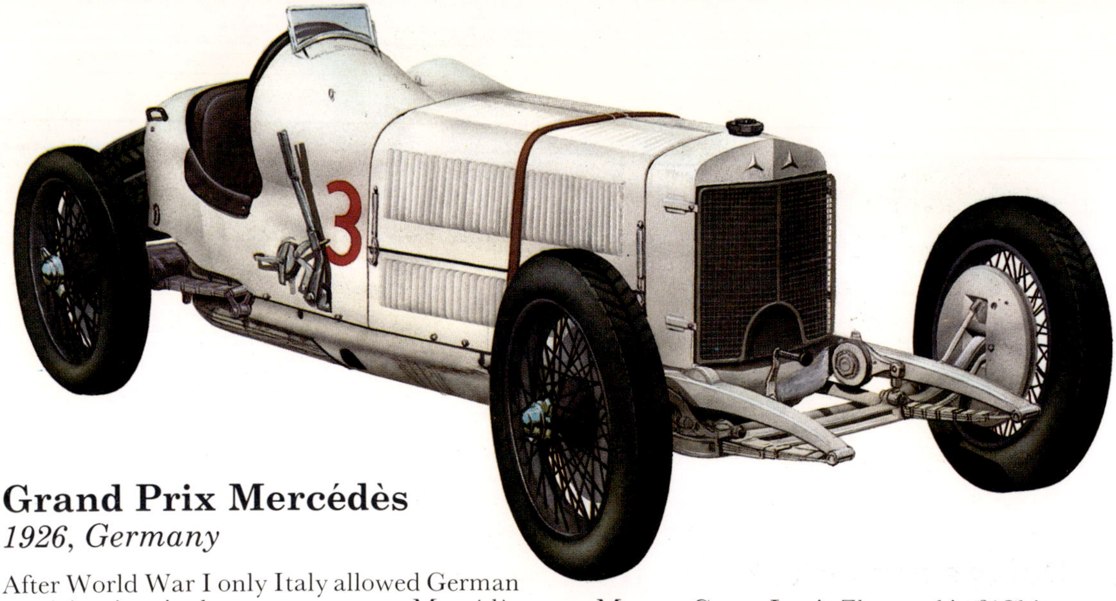

Grand Prix Mercédès

1926, Germany

After World War I only Italy allowed German participation in her motor races. Mercédès took their team of cars there in 1922 for the Targa Florio–including two pre-war 4.5-litre Grand Prix models for Lautenschlager and Salzer, Count Masetti's privately-owned example, two supercharged 28/95s for Sailer and Werner and a pair of new 1.5-litre four-cylinder supercharged cars for Scheef and Minoia. Masetti won the race outright. Mercédès were obviously still very effective.

The company progressed through many minor events until 1924 when Ferdinand Porsche designed for them a straight-eight Grand Prix car of 1980 cc with twin overhead camshafts, four valves per cylinder, sodium-cooled exhaust valves, nine-main-bearing crankshaft and dry-sump lubrication–nearly all features which breathed 'racing' for many many years to come. These engines delivered about 160 bhp at over 7000 rpm but were installed in wickedly handling chassis which won the cars a reputation as driver-killers.

In the cars' debut in the Italian Grand Prix at Monza, Count Louis Zborowski of 'Chitty-Bang-Bang' fame was killed in one when he lost control and hit a tree. In 1925 Werner used one of the cars to achieve fastest time at the tricky Freiburg Mountain Climb, proving it could be tamed. In 1926 they won the Freiburg, Solitude and Klausen climbs. At the last-named venue a young driver named Rudolf Caracciola ('Karachola') brought his difficult K-type Mercédès to the summit in a time very close to that of works driver Rosenberger in the Grand Prix model. For the inaugural German Grand Prix race on the AVUS circuit at Berlin Caracciola was given a straight-eight and he won with it brilliantly in driving rain.

Yet another bad accident in that race involved a sister car, and in 1927 the Porsche-designed straight-eight was slowly phased out. It gains its place in history for providing Caracciola with the first of his many Grand Prix victories, and for proving Mercédès' postwar engineering skills, despite their mistakes in its chassis design.

Rudi Caracciola and his mechanic acknowledge the plaudits of the crowd after winning the first German Grand Prix, in 1926 at AVUS (Berlin) in the wicked straight-eight two-litre Mercedes.

Fiat Tipo '806 Corsa'

1927, Italy

Having withdrawn from full-time motor racing in pique at having so many of their top engineers stolen by Alfa Romeo and Sunbeam, the Fiat company in Turin set about a return to the sport for promotional purposes in 1926–7. Two notable engineers, Luigi Cavalli and Tranquillo Zerbi, had always remained loyal. When barely 30, Zerbi had been largely responsible for the first supercharged Fiat engine and during 1925 he began work with engineers Sola and Treves on a design for the 1500 cc Grand Prix Formula which was to run through 1926–27. They were attracted by the promise of two-stroke engine power but their advanced test engine exploded repeatedly.

So their back-up design was eventually brought out and dusted-off. This design, which became the startling Fiat 806, was for a twelve-cylinder engine arranged virtually as two straight-six-cylinder units, standing side by side in the front of the car with their individual crankshafts geared together. There were three overhead camshafts to control inclined valves, the common centre camshaft actuating the inlet valves for both six-cylinder units. Bore and stroke were 50 mm and 63 mm, for 1484 cc; it weighed only 381 lbs thanks to extensive use of light-alloy construction, and delivered 187 bhp at 8500 rpm – far more than the 500 lb 170 bhp Delage.

The engine was installed in a wickedly low chassis, with rakish Grand Prix offset-single-seat bodywork and, after initial tests showed the power unit to be fragile, its tune was reduced to give 160 bhp at 8000 rpm.

Cavalli and Zerbi wanted to run the car in the 1927 Italian Grand Prix but Agnelli was not keen to place Fiat's reputation on the 806's unproven shoulders. Bordino was entered in the car in a supporting race at Monza, and won brilliantly, lapping far faster than Benoist's Grand Prix winning Delage. The 806 never raced again, having proved Fiat's point.

Grand Prix Talbot

1927, France

The Suresnes-built Talbots were a product of the confusing Anglo-French Sunbeam-Talbot-Darracq combine during the 'tween-war years, and as profit-conscious shareholders in the British Sunbeam wing of the company would not countenance continued Sunbeam competition programmes in 1926–27, it fell to Talbot to contest the 1.5-litre Formula.

Louis Coatalen was still in charge, with ex-Fiat engineers Bertarione and Becchia responsible for design. In some ways the Talbot Grand Prix cars for the new season were even more advanced than Albert Lory's quite extraordinary Delages. Their straight-eight engines had pairs of forged steel cylinders with welded-on cooling-water jackets and roller-bearing crankshaft, and was offset to the left of the car so that the driver could be placed low down on the undertray, beside the propeller-shaft. The front axle was made in two straight bored-out sections with the springs passing clean through it, the halves being bolted together on the centre-line. What really set the Talbot apart from the Delage was its chassis, which consisted of strong box-members spaced vertically ten inches apart with robust struts welded between them. In addition the whole structure was stiffened by screwing the body panels rigidly to the side members. With over 140 bhp at 7000 rpm, these very light and rigid supercharged cars offered extremely good handling characteristics if their drivers could adjust them properly. Potentially they had the legs of a Delage, despite their power deficit.

Unfortunately front axle failures sidelined the cars in their debut in the 1926 British Grand Prix at Brooklands, but Segrave won the JCC '200' there and the cars were placed 1–2–3 against meagre opposition at Montl-

héry. They did little better in 1927 as the money was running out, and from 1928 to 1930 they raced as Italian private entries. The Talbot Grand Prix car was an outstanding 'might-have-been'.

Maserati 'Sedici Cilindri'
1929, Italy

After the demise of the $1\frac{1}{2}$-litre Grand Prix Formula at the end of 1927 virtual anarchy replaced it during the Depression years up to 1934. It was during this 'Formule Libre' period that a breed of new road-racing specials appeared; those manufacturers who could afford the luxury of motor-racing involvement put the largest possible engines into whatever sketchy chassis frame they could confect.

On July 1, 1929, a new car built by the Maserati brothers in Bologna appeared at Cremona in Italy for a race on the very fast 25-mile public road course there. Baconin Borzacchini – now there's a name to conjure with! – was entrusted with the car and he howled through a 10 km timed section of the course to average 152.9 mph – the highest speed ever recorded on a road circuit at that time. The 'Sedici Cilindri' (sixteen-cylinder) Maserati had proved itself the fastest road racing car in the world.

The brothers, Alfieri, Bindo, Ettore and Ernesto, had taken two of their two-litre eight-cylinder engines and mounted them upon a common light-alloy coupled crankcase. Bore and stroke were 62 mm and 82 mm respectively to give 3961 cc. The brothers claimed 305 bhp at 5200 rpm from their monster, which they called the Maserati V4. Alfieri set a lap record of 124.2 mph at Monza and in 1930 Borzacchini won with V4 at Tripoli. It ran poorly without its twin superchargers at Indianapolis and Ernesto shattered the Pescara lap record with them replaced.

In 1931 Ernesto won the Rome Grand Prix with the ageing monster and for 1932 a larger version, V5, was constructed with 4905 cc and over 330 bhp. Luigi Fagioli set a new Monza lap record of 112.2 mph with this car on the revised Milanese circuit and then won the Rome Grand Prix with it. René Dreyfus set a 130.87 mph lap record at AVUS and Fagioli won at Monza. At the end of the year a driver named Ruggeri died in V5 during a record attempt at Montlhéry. During 1933 V5 was slowly rebuilt; with a more modern body style and 350 bhp it reappeared at Tripoli in 1934, driven by Piero Taruffi. He crashed heavily, ending the car's career.

Old V4, rebuilt as a sixteen-cylinder road car, survives today.

The Miller 91s won almost everything in America in the late twenties. A front-drive car is shown here.

Miller 91

1929, United States

Harry Armenius Miller was a towering figure in American motor racing between the wars.

Miller built carburettors originally, and turned out his first engine in 1916, but it was after World War I that the Miller marque came to prominence. He worked with draughtsman Leo Goossen and works manager Fred Offenhauser to develop the best ideas then current in racing. They drew on Duesenberg's straight-eight engine layout, Peugeot's inclined four-valve-per-cylinder head and twin overhead camshafts, and Ballot's piston-type cam-followers, and produced a three-litre straight-eight of their own which was very advanced and powerful. With this engine in a Duesenberg chassis Jimmy Murphy won the 1922 Indy 500-Miles.

In 1923, 122-cubic inch regulations were adopted to form a two-litre Formula for American racing.

Nine new Millers started at Indianapolis, and four of them finished 1–2–3–4. Miller dominated the rest of the 1923 US track season and although losing Indianapolis to Duesenberg in 1924, won virtually everywhere else.

In 1925 Miller produced his jewel-like front-wheel-drive model which led a Miller 2–4–5–6–7–9 result at Indianapolis. These cars were incredibly fast, Millers taking AAA records at 141.2 mph.

For 1926 a new 91-cubic inch (1491 cc) Formula was applied and the Miller '91' was built to comply. It was breathtakingly beautiful, a true little single-seater with supercharged engine giving 154 bhp at 7000 rpm, and later, tuned by Frank Lockhart, 285 bhp at 8000 rpm. Lockhart won Indy.

Miller charged $15,000 for front-drive cars, $10,000 for rear-drives; 50 '91s' were built of which perhaps only 12 had front-drive. While the front-drive was faster in a straight line the rear-drive had a better racing record, winning Indy again in 1928 and 1929, while in 1927 Lockhart – later killed attempting the Land Speed Record in a car of his own design – covered a flying mile at 164 mph average. The Millers won almost everything in America in the late twenties, and when showman driver Leon Duray took a pair to Italy for the Grand Prix of 1929 and also ran them at Montlhéry near Paris, Ettore Bugatti took close interest, acquired both cars and adapted the twin-cam head for his own outstanding motor cars.

Alfa Romeo Tipo A 'Bimotore'

1931, Italy

Alfa Romeo were intrigued by Maserati's display of speed and power – though coupled to the usual Bologna unreliability – with the Sedici Cilindri V4, and for 1931 Vittorio Jano drew up a fearsome Formule Libre Grand Prix car of his own, which became known as the Tipo A.

Jano took two of the company's existing and very successful 1750 cc supercharged six-cylinder engines and mounted them side-by-side in a style similar to the Maserati V4. The engines were 'handed' so that their exhausts exited respectively on the left and right side of the car, and their crankshafts revolved in opposite directions to counter one another's torque reaction. On the rear of each power unit was a gearbox, with both individual clutches coupled to the same pedal, and there were also two gearlevers, interlinked so that the driver could change with either hand. Two parallel propeller shafts ran to individual bevel-gear final drives for each rear wheel.

The car was completed with the first genuine '*monoposto*' (centrally-disposed single-seat) bodyshell seen in a European Grand Prix racing car. The engine dimensions were 65 mm × 88 mm, giving a total twelve-cylinder capacity of 3504 cc and power output of about 220 bhp at 5000 rpm, although at least 300 bhp was claimed at the time.

Grand Prix regulations then asked little more than that race endurance should be not less than a staggering ten hours and on May 24 the new Tipo A made its debut in the Italian Grand Prix at Monza. Unfortunately, Luigi Arcangeli was killed in the car during practice, but the car was repaired and driven by Nuvolari and Borzacchini in the race – where it proved a pig to handle and flopped. Later that year Giuseppe Campari won the important Coppa Acerbo race at Pescara in the car, now better developed and more manageable, and for the Monza Grand Prix two Tipo As were entered for Campari and Nuvolari. Both failed, one with gearbox and the other with tyre troubles and the double-six Alfa was never to be raced again. Even so, they had showed Jano the way ahead.

Bugatti Type 51

1931, France

The stimulus of Maserati did more than merely fire Alfa Romeo into action to replace their obsolescent P2 model, for it crossed the border into France and triggered Ettore Bugatti.

He realized that to maintain his marque's position in the high-performance hierarchy he would have to adopt twin overhead camshafts, and perhaps also subscribe to the cubic-inch theory – 'there's no substitoot for coobic inches' – by building a large-capacity racer in the manner of the Sedici Cilindri Maserati.

'Leon Duray' had sold his two front-drive Miller 91s to Bugatti in part-exchange for three of the Molsheim cars. Thus Bugatti won himself a highly-developed inclined-valve cylinder head with twin overhead camshafts which he adapted first to his 86 mm-bore 5.3-litre Type 46 model in 1930, and later with very little alteration to the 60 mm-bore 2.3-litre Type 51 for 1931.

In both cases the cylinder head was one with the block, but the 2.3 Type 51 showed 187 bhp at 5200 rpm compared to the conventional 2.3 Type 35T's output of about 135 bhp. Bugatti built Type 51s in series, and they completely re-established Molsheim's supremacy during 1931 by winning no less than 11 of the 18 major events held.

Louis Chiron won the Monaco and Czechoslovakian Grands Prix in 51s; he shared victory at Montlhéry with Achille Varzi; 'Williams' and Conelli won at Spa in Belgium; Varzi won at Tunis; the Polish Count Czaykowski won in Morocco . . . so the list goes on.

The Type 51 looked every inch the true Bugatti it was, with that Type 35-like body design, combining the Grand Prix and high-performance sporting traditions, and cast-alloy wheels integral with the brake drums. In the ten-hour Grand Prix races of the age these contributed materially to success, saving much time during pit stops when brake linings as well as tyres had to be changed. 'Two-three' twin-cam Bugattis raced on all over the world and many Type 51s are still extremely robust and healthy, winning club races to this day.

GP Bugatti with twin overhead camshafts, developed from the Miller 91.

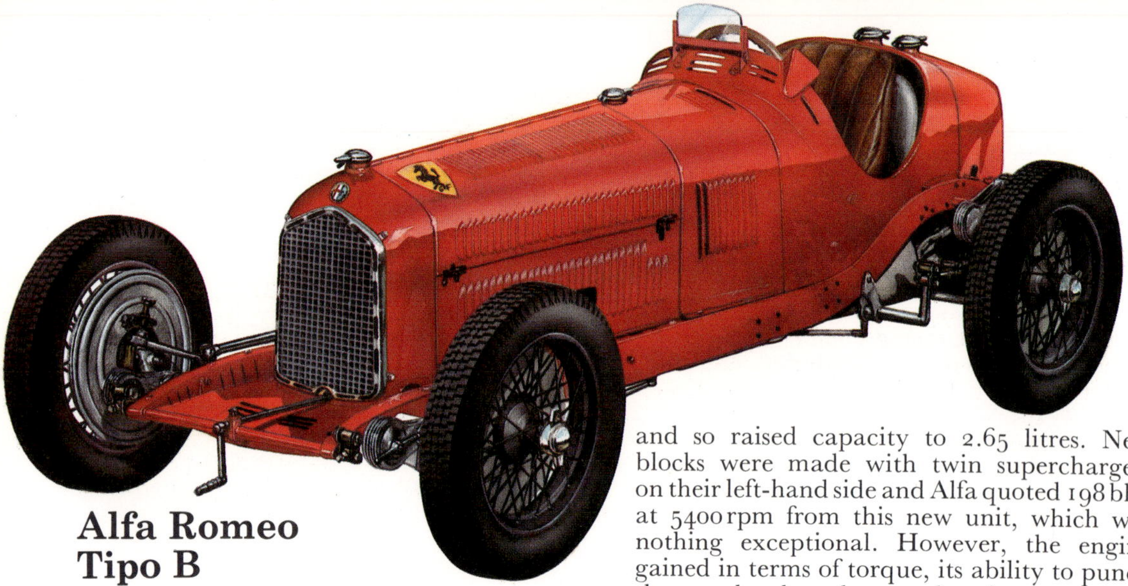

Alfa Romeo Tipo B

1932, Italy

After the lessons learned with the fearsome Tipo A *monoposto*–literally 'single-seat'–racing car in 1931, Vittorio Jano of the Alfa Romeo factory at Portello, Milan, began development work on a less complex car which would offer little less power but a great saving in weight for 1932. If a new car was to be laid down for Grand Prix racing without thought of turning it into a production model, there was no logical reason to make it anything other than a true single-seater–like the Tipo A special.

Alfa Romeo had generally been racing a two-seat-width 2.3-litre straight-eight supercharged model known as the 'Monza'. Now Jano took this proven engine, added 6 mm to the crankthrow to give a bore and stroke of 65 mm × 100 mm instead of 65 mm × 88 mm, and so raised capacity to 2.65 litres. New blocks were made with twin superchargers on their left-hand side and Alfa quoted 198 bhp at 5400 rpm from this new unit, which was nothing exceptional. However, the engine gained in terms of torque, its ability to punch the car hard under acceleration, and in a narrow ladder chassis frame and single-seat body of modest weight it was to prove very effective.

Perhaps the most unusual aspect of what in truth was an economy racing car–for Alfa's production at the time was merely ten road cars a week–was Jano's rear suspension arrangement in which the drive was divided along two individual propeller-shafts, one bevel-driving each rear wheel from a central differential mounted on the tail of the gearbox. The reason for this layout has long exercized historians' minds, but probably it was intended to minimize the tendency of the rear axle to hop under acceleration and so cause wheel-spin.

▼ Monopostos *in line astern during the 1935 Dieppe Grand Prix with private owners Dick Shuttleworth, and 'Georges Raph' leading René Dreyfus (temporarily).*

The slim new body was only 26 inches wide with low-cut cockpit sides to give the driver elbow room. The new model – instantly dubbed the P3 by journalists – made its debut in the then five-hour Italian Grand Prix at Monza in June 1932 and it won, driven by Nuvolari. Three more major victories fell to the model that year and then Alfa Romeo bowed to economic strictures at the season's end and left their racing honour in the hands of Enzo Ferrari's private Scuderia, using older models, while the new Tipo Bs were placed under wraps.

When the re-engined Monzas campaigned by Ferrari proved unreliable the Milan management relented and released the *monoposti*. Fagioli drove the first car to victory at Pescara, and he and Louis Chiron rapidly added four more Alfa Romeo victories.

In 1934 the new 750kg maximum-weight Formula took effect to regularize Grand Prix racing. The German state-backed firms of Mercedes-Benz and Auto Union produced advanced all-independently suspended, supremely powerful racing cars which should have made the solid-axle cart-spring P3s obsolete overnight. An 85cm minimum-body-width rule led to the *monoposto* cockpit sides being bulged and with 2905cc engines and 255bhp they won no less than 13 major events, old 2.6 models winning twice and a streamlined 3.2 litre special once at AVUS.

For 1935 modified suspension, 3165cc and later 3822cc power units were adopted to stave off the German horde, but the ageing, by this time immortal cars were generally only successful in the absence of Mercedes-Benz and Auto Union. Nuvolari won the German Grand Prix for Alfa Romeo to grind the German nose into the soil, and the Tipo Bs then became virtually the first recognized historic racing cars. They were the first significant Grand Prix cars to use single-seat, slimline, slipper bodies – the first of a noble breed.

▲*For 1934 and the start of the 750kg maximum weight regulations a new minimum cockpit width regulation was included. The Alfa* Monoposto's *slim sides had to be bulged to meet it.*

Monoposto *Alfa Romeos prominent on the Monaco Grand Prix grid, Monte Carlo in 1932.*

Gentle giant John Cobb in the cockpit of his Napier-Railton with co-drivers Tim Rose-Richards (left) and Charlie Dodson at Brooklands, 1933.

Napier-Railton Special
1933, Great Britain

John Cobb was a giant of a man, tall and heavily-built, yet extremely modest by nature, very quiet and retiring. He was a wealthy furrier, and his great enthusiasm was for large-engined cars and for racing them on the Brooklands Outer Circuit. For 1933 he contracted Thomson & Taylor of Brooklands to build a very special car for him, using a surplus 23,970cc 505bhp Napier Lion aero engine. This unit was unusual in disposing its twelve-cylinders in three banks of four–what was known as a 'broad-arrow' formation.

Reid Railton designed a massive chassis to accommodate this very powerful yet relatively lightweight engine, using a single-seater body and dispensing with details like front-wheel brakes. One prime target was the world's 24-Hour record, but in 1933 Cobb began his new car's career by setting a 120.59mph standing-start lap record on his beloved Brooklands Outer Circuit.

The first attack on the 24-Hour record failed at Montlhéry in 1934 when Freddie Dixon crashed the car, although in the same year Cobb twice broke the Outer Circuit lap record at Brooklands. In 1935 Cobb, Tim Rose-Richards and Dixon took the 24-Hour record at 137.40mph and at Bonneville Salt Flats in America they also took the Hour record at 152.7mph. The car spent long hours above 165mph and back home Cobb drove it to win the Brooklands 500-Miles at 121.28 mph. In October 1935 he raised the Outer Circuit lap record yet again, to 143.44mph, and that figure was never bettered. In 1936 Cobb and the Napier-Railton raised the 24-Hour record to a stunning 150.6mph and the old car made its swansong appearance in Cobb's hands in 1937, when it won the BRDC 500 once again at Brooklands.

John Cobb went on to take the World Land Speed Record in his twin-engined Railton-Mobil Special. His many friends were grief-stricken when he lost his life on Loch Ness in 1952–his Water Speed Record attempt in the jet-boat *Crusader* ended in the craft's disintegration. The Napier-Railton is still around today, in superb condition.

Auto Union A-Type
1934, Germany

Auto Union was formed by businessman Jorgen Skafte Rasmussen from the German motor companies Audi, Horch, Wanderer and DKW. In an effort to promote their image as high-grade automotive technicians, Auto Union considered building cars for the new 750kg Formula due to take effect for 1934–36. When Ferdinand Porsche approached them with a startling design reminiscent of the ten-year-old Benz *Tropfenwagen*, they said 'yes please' and set about building the cars using

in part a German Government grant authorized by racing-enthusiast Adolf Hitler.

The Auto Union design was unusual in placing the engine between the driver and the car's rear axle, for its V16-cylinder engine layout and for all-independent suspension, which had swing axles *à la* Benz at the rear. The 4.4-litre supercharged but simple (single central camshaft) engine delivered around 295 bhp at 4500 rpm, but with immense torque. The car was clad in a long-tailed teardrop bodyshell.

Early in 1934 Hans Stuck set a class Hour record of 134.9 mph at AVUS and the team was placed third in its first race, again at the Berlin track. Auto Union won the year's German Grand Prix to widespread delight and the Swiss and Czechoslovak Grands Prix too – Stuck humbling Mercedes-Benz in this revolutionary racing car.

For 1935 the Auto Union B-Type appeared, with 4.9-litre 370 bhp V16 engine, Stuck winning the Italian Grand Prix and Varzi the Tunis Grand Prix and the Coppa Acerbo; while ex-motorcyclist Bernd Rosemeyer won the Czech race. The Auto Unions could not match Mercedes this season, their swing-axle rear suspension and immense power making them tricky to handle. However, Rosemeyer was an instinctive driving genius and he tamed the cars better than most. In 1936 he swept all before him, winning the Eifel, German, Swiss and Italian Grands Prix to become European Champion – in effect World Champion of the day. His C-Type car had a six-litre V16 engine giving 520 bhp and when the 750 kg Formula was extended through 1937 Rosemeyer won the Eifel, Pescara, Donington and Vanderbilt Cup races; team-mates Rudi Hasse and Ernst von Delius took the Belgian Grand Prix and a South African race. From the pen of Dr Porsche had come a fantastic family of sixteen-cylinder cars.

Teutonic fury; the 1937 six-litre supercharged V16-cylinder Auto Union C-Type was fiercely functional.

H. P. Muller yumps his C-Type Auto Union at Donington during the legendary 1937 Grand Prix at the Derby circuit in England.

Supercharging arrangements clearly visible in the W25.

Mercedes-Benz W125

1937, Germany

If Auto Union staggered the motor racing world in the thirties by innovation and unconventionality, then Mercedes-Benz refined conventionality to a crushing degree. Mercedes-Benz were also backed financially by the Third Reich, but this Government subsidy went only a little way towards covering the vast expenditure which both German teams made – Mercedes-Benz much more so than Auto Union.

For the start of the new Formula in 1934 design director Dr Nibel laid down the twin-overhead-camshaft W25 straight-eight engine with four valves per cylinder and supercharging, displacing 3.36 litres with bore and stroke dimensions of 78 mm and 88 mm. This unit weighed 450 lb and was supercharged through a system which incorporated a bypass to exhaust blower pressure to the atmosphere when the driver shut the throttle. This caused an ear-splitting shriek which became typical of the cars. While the enlarged Alfa Romeo Tipo Bs had 19.5 bhp per square foot of frontal area and 235 bhp per ton on the start-line, the Auto Union A-Type had 24 bhp per square foot and 260 bhp per ton, and the Mercedes-Benz 30 bhp per square foot and 305 bhp per ton – the German car's modern styling and advanced metallurgy aggravating the effects of Italian financial stringency.

Through that first season the W25's engine size rose to achieve 398 bhp; Fagioli won the Coppa Acerbo and Spanish Grands Prix and shared the Italian Grand Prix victory with Caracciola. For 1935 four-litre engines were standardized and W25s won the Eifel, Tripoli, Monaco, AVUS, French, Swiss, Spanish and Belgian events.

Things went badly wrong in 1936 as Mercedes adopted more powerful 4.7-litre 474 bhp engines in short wheelbase chassis and the result was unmanageable handling reminiscent of the 1924 car at its worst. Caracciola won at Monaco where the tight circuit suited the tail-happy cars, and at Tunis; while that winter saw the driver from Remagen take a new streamlined version with 5.5-litre V12 engine over the mile at 228.05 mph.

For the extra year of 750 kg racing in 1937 the technical office at Stuttgart-Untertürkheim turned out the epochal W125 model, which with a 5.66-litre supercharged straight-eight engine achieved the greatest power output ever seen in Grand Prix racing – 646 bhp at 5800 rpm, a figure yet to be surpassed. In addition a new twin-tube chassis frame was adopted, plus redesigned wishbone front suspension. Two wishbone-shaped components attached the wheel hub assembly to the chassis, controlled by a coil-spring inserted between them. At the rear a De Dion axle layout was adopted. This approach dates from the dawn of motoring in fact, but it is in truth a very practical system: the wheels are connected by a rigid beam serving to keep them parallel at all times in the vertical sense, so resisting the up-edging effect of swing axles whereby a tyre can lean over so much during cornering that its contact patch area with the roadway is drastically reduced.

During 1937 the fearsome W125s proved the ultimate development in 750 kg Formula racing, capable of spinning their rear wheels at 170 mph or more in top gear. Ex-mechanic Hermann Lang won at Tripoli, and in a streamliner at AVUS – lapping at 171.63 mph there; Caracciola won the German Grand Prix, Brauchitsch won at Monaco and Caracciola again at Livorno.

The W125s were retained for hillclimb work during the three-litre Grand Prix Formula years of 1938–39 and after the war became prized museum pieces. Outside Mercedes-Benz, one survives today in racing condition, owned by Neil Corner in Britain, and it is utterly beyond price.

Hermann Lang demonstrates the factory's W125 at Montreux in 1978.

Manfred von Brauchitsch in the original 1934 Mercedes W25 Grand Prix car tackling the Kesselberg mountain climb.

ERA

1935, Great Britain

During all these years of development by the Americans, French, Italians and Germans, and since the demise of Sunbeam, Britain had been nowhere on the international racing scene at top level.

Up at Bourne in Lincolnshire, Raymond Mays, son of a local wool-broker, had pursued a long and very successful career of sprints and hill-climbing with occasional races at Brooklands thrown in for good measure. In 1933 he obtained backing from wealthy businessman Humphrey Cook to develop with Peter Berthon a much-modified Riley design they had put together, for use as a true single-seater in international Voiturette or 'Formula 2' racing.

Thus was born the English Racing Automobile, and after conquering minor teething troubles early in 1934 the six-cylinder 1100 cc and 1500 cc ERAs proved supremely effective. Peter Berthon worked with supercharger expert Murray Jamieson and chassis designer Reid Railton in building the cars.

Only 17 of the basic uprights were built and they raced throughout Europe and in Australia, South Africa and America in the middle and late-thirties. In 1937 ERA was the top-scoring racing make with 14 victories to Mercedes-Benz's seven, although of course the German team achieved theirs in far more august events.

Output of the original 1.5-litre engine was about 150 bhp at 6500 rpm and Mays' personal two-litre model was subsequently developed with Zoller supercharging to achieve as much as 340 bhp. The cars were owned and driven by such famous names as Dick Seaman, Prince Birabongse of Siam – racing under the pseudonym 'B.Bira', Pat Fairfield, Charlie Martin, Peter Whitehead, Peter Walker, Arthur Dobson, Lord Howe and Marcel Lehoux – the last unfortunately losing his life in the only ERA to be totally written off.

Alfa Romeo 12C-36

1936, Italy

In the late thirties once so dominant Italian companies such as Alfa Romeo fought a desperate rearguard action to stem the advance of the better-financed and always more organized German GP teams.

Once it had become apparent that the great Tipo B Monoposto cars were utterly obsolescent, their experimental 3822 cc, 78 mm × 100 mm straight-eight supercharged engines – developing around 330 bhp at 5400 rpm – were installed in modernized all-independently-suspended chassis frames with bulbous, streamlined, slipper bodies similar to those in use by Mercedes-Benz. The result was the 8C-35 Alfa Romeo Grand Prix model, but sadly any improvement the car achieved in handling over the Tipo B still left it barely on a level with the German cars, and of course it lagged way behind them in sheer power.

So a new V12-cylinder engine was developed at Portello and in the shops of the quasi-works Scuderia Ferrari team at Modena. This power unit had dimensions of 70 mm × 88 mm and a displacement of 4064 cc, achieving a claimed output in supercharged form of 370 bhp at 5800 rpm with a considerable increase in torque, smoothness and general flexibility. In practice the improvement was not noticeable.

However, the driving skill of Tazio Nuvolari was always worth several tens of horsepower and in 1936 his 12C beat the Germans in the Penya Rhin Grand Prix at Barcelona, and again at Budapest and in the Milan Grand Prix around Sempione Park. Nuvolari also won the 1936 Vanderbilt Cup in America, using an artificial track laid out around Long Island's Roosevelt Field airport. In 1937 Nuvolari's 12C-36 beat Hasse's Auto Union

in Sempione Park, but fourth place in the German Grand Prix was the best in a major race that year.

Delahaye Type 145 V12
1938, France

If the Italians took a bruising in the late thirties at the hands of the German state-backed teams, then the poor French were utterly pounded into submission. Bugatti steadfastly refused to adopt such trickery as all-independent suspension and virtually faded from the racing scene. Whenever possible the French ran their major races around tightly-chicaned circuits unsuitable for the German cars' speed and power, but when this ruse failed they turned to sports-car racing.

For 1938 new Grand Prix regulations took effect. In place of the 750 kg maximum weight Formula of preceding years, there was now a minimum weight limit of 850 kg for cars limited to three litres capacity supercharged, or $4\frac{1}{2}$ litres unsupercharged. It was apparent that the allowance for unsupercharged engines was inadequate but nevertheless the French Talbot and Delahaye concerns took the plunge.

Hey presto – in the first race of the new Formula at Pau in 1938, René Dreyfus won for the Delahaye truck company, beating the latest Mercedes-Benz on economy. Designer Jean François had produced for Delahaye the Type 45, an ugly V12 of 4482 cc with pushrod-operated valves, De Dion rear axle and four-speed Cotal gearbox. In original trim this truck-like grotesque was to have been a two-seater useable on the road, but in *monoplace* Grand Prix form it achieved around 238 bhp from its unblown $4\frac{1}{2}$ litres and could reach over 160 mph with a long enough run-up. Dreyfus won in the car at Cork, Eire, in 1938, but it was totally outclassed when it faced the Germans again in Switzerland, Germany, and Tripoli and at Donington Park in England. During 1939 the car was utterly outclassed again, although Dreyfus and Varet brought a road-equipped version – fitted with wings and lights – in fourth in the Mille Miglia.

The 1938 Alfa 158, conceived by Enzo Ferrari, and forerunner of one of the great Grand Prix cars of all time. Its postwar descendant is a legend . . .

Alfa Romeo 158

1938, Italy

Enzo Ferrari had been virtually in charge of Alfa Romeo's entire racing operation during the thirties, under the auspices of his private Scuderia Ferrari. This was founded as a commercial self-help concern in 1929 to prepare and maintain Alfa Romeo competition cars – and for a short period other manufacturers' racing motor-cycles – for a circle of wealthy private owners.

Continual failure in Grand Prix racing, however, was doing neither Alfa Romeo nor the Scuderia Ferrari any good. To counter some of Maserati's burgeoning prestige within Italy, Ferrari himself founded a project to produce a 1.5-litre voiturette racing car for the 1938 season.

Gioacchino Columbo – a former assistant of the great Jano – was given the job, and he evolved a 1.5-litre straight-eight engine which was in effect one half of Alfa's three-litre Grand Prix 316 V16-cylinder. However, the gear-train to drive the twin overhead camshafts and other ancillaries on the new engine was placed at its front, unlike the V16 which had the train at its rear and previous Alfa 8s which had it between cylinders numbers four and five.

On the test bed at Modena the new engine gave 180 bhp at 6500 rpm, and by the time of the new car's racing debut at Livorno in 1938 had reached 195 bhp at 7000 rpm. Emilio Villoresi was the driver at Livorno, and he won again in the Milan Grand Prix; but elsewhere the cars failed late that year. Main-bearing lubrication was improved during the winter and in 1939 the 158s offered 225 bhp at 7500 rpm. The Tripoli Grand Prix began the new season, but Mercedes-Benz sprang their secret 1.5-litre V8 W165 model on the shocked Italians and beat them hollow . . . The Alfas retired with overheating in the desert sun, but Farina won at Livorno, Biondetti at Pescara and Farina again in the Prix de Berne. In the Swiss Grand Prix Farina ran second in the rain amongst the German Grand Prix cars!

Racing continued in Italy in 1940, Farina winning at Tripoli, but that winter Emilio Villoresi was killed testing at Monza and promising young Aldrighetti had crashed fatally in a 158 at Pescara. The Alfetta was dominant, but at a price, and in the absence of the Germans.

Mercedes-Benz W154

1938, Germany

With the change in Formula for 1938–39 both Mercedes-Benz and Auto Union developed V12-cylinder three-litre supercharged engines; the Stuttgart engineers produced a unit with twin overhead camshafts per bank, while Zwickau adopted three-cam design with a single, overhead, exhaust-valve cam-

shaft on each bank of six-cylinders and a single, central, inlet-valve camshaft doubling-up for both banks amidships within the engine 'V'.

While Mercedes achieved 420 bhp from their V12, having spared no expense in its design and construction, Auto Union were operating on a very tight budget – even to the extent of using the ex-750 kg Formula cars' gearboxes for their new D-Type three-litre model, and their engine offered about 400 bhp. One major change in the Auto Union was in the chassis design, whereby a De Dion system replaced the swingaxles at the rear.

Through 1938 the Mercedes-Benz cars were dominant, following their initial surprise defeat at Pau, in France. Auto Union missed races while completing their development, and were reeling from the loss of Rosemeyer in a record-attempt that winter; Mercedes' W 154 models were therefore placed 1–2–3 at Tripoli, Dick Seaman won the German Grand Prix, Lang won in Italy and so on.

Auto Union finally appeared with their neat new cars and Nuvolari to take Rosemeyer's place; the Italian master won the Italian and Donington Grands Prix for the team. But in 1939 Mercedes adopted two-stage supercharging, in which a small compressor fed a larger compressor and then the total boost was applied to the engine; the V12 developed 480 bhp in this form and Lang was dominant with the W 154/163 rebodied model. Seaman died in his car in the Belgian Grand Prix while leading in the rain. Auto Union adopted two-stage supercharging in time for the 1939 French Grand Prix at Rheims, H. P. Muller winning for them, and Nuvolari drove to win the last pre-war Grand Prix of all, at Belgrade on the day war was declared between Germany and Great Britain. Mercedes-Benz, who by the end of 1939 had been approaching 1937 lap speeds, revived their 1939 cars to race in Argentina in 1952.

Donington Grand Prix 1938 with Muller's Auto Union D-Type V12 holding off Von Brauchitsch's Mercedes-Benz W154 V12 – mid-engine design versus front-engine.

Maserati 4CL

1939, Italy

Maserati found that 1.5-litre voiturette racing was a much more sensible proposition for a factory such as theirs during the thirties. In the face of ERA opposition from Britain they eventually achieved a dominance in that class and a degree of Italian national prestige which led Alfa Romeo once again to react to the Maserati brothers' stimulus and turn to building a new voiturette themselves. Since the Italians could not play in the same park as the big boys – the German teams – they took their ball away . . .

Maserati introduced their six-cylinder 6C model for voiturette racing in 1936, replacing the formal four- and eight-cylinder models. In general design it followed on from the eight-cylinders but the twin-overhead-camshaft 1493 cc power unit was good for 155 bhp at a smooth 6200 rpm. From 1936 to 1938 the Maserati 6C was very competitive with the similarly six-cylinder ERA and in Italian national events and most others around Europe it numerically saturated the field.

It was in response to the success and prestige achieved by these 6Cs that Alfa Romeo weighed in with their straight-eight Tipo 158 voiturette in 1938; to combat its threat the ever-cheerful and willing Maserati brothers produced in turn their 4C, using the 6C chassis to carry 1088 and 1496 cc twin-overhead-camshaft engines. Still the Alfetta proved to be streets ahead, and in 1939 the 4CL was released with four valves per cylinder.

As the 'Sixteen-valve' Maserati this model became famous, as did its 'square' dimensions of 78 mm × 78 mm to give 1489 cc, and its supercharged power output of 220 bhp. However, while the Maserati won everywhere not attended by the Alfetta team, it could not combat the 158s on equal terms. But postwar, Maserati would be back, with modified versions of this late-thirties voiturette . . .

'Boyle Special' Maserati 8CTF

1939, Italy

Maserati actually led the German Grand Prix of 1939 for a time, the three-litre supercharged straight-eight driven by Paul Pietsch sorting out the German cars on the opening lap. But fiercely accelerated and braked as is necessary on a road circuit, it never had the stamina to maintain the lead. The high-speed constant grind of the Indianapolis 500-Miles track race, with left-turns only over a $2\frac{1}{2}$-mile rectangular course containing four banked 90-degree corners, was a different type of challenge; here the 8CTF was to shine.

The brothers had built the 8CTF using virtually two 4CL sixteen-valve four-cylinder 1.5-litre engines coupled in tandem to form a 2990 cc unit giving about 355 bhp. One car was bought by Americans and ran at Indianapolis in 1939 as the Boyle Special, being driven by the local ace Wilbur Shaw. The car set new performance standards at the Speedway, with its independent front suspension and well-located live rear axle, and Shaw won handsomely. In 1940 he returned with the same car and won again, with similar ease. Since the USA was still not embroiled in World War II, there was another Indy 500 race in May 1941, and Shaw was all set to achieve a staggering hat-trick of victories driving Boyle's famous car. Before the race Shaw had some misgivings, and sure enough he lost the lead when a wire wheel broke its spokes and collapsed; the leading Maserati spun heavily into a retaining wall and damaged Shaw's spine.

After the war this remarkable old car was revived and raced again at Indy by Ted Horn. He placed it third in both 1946 and 1947, by

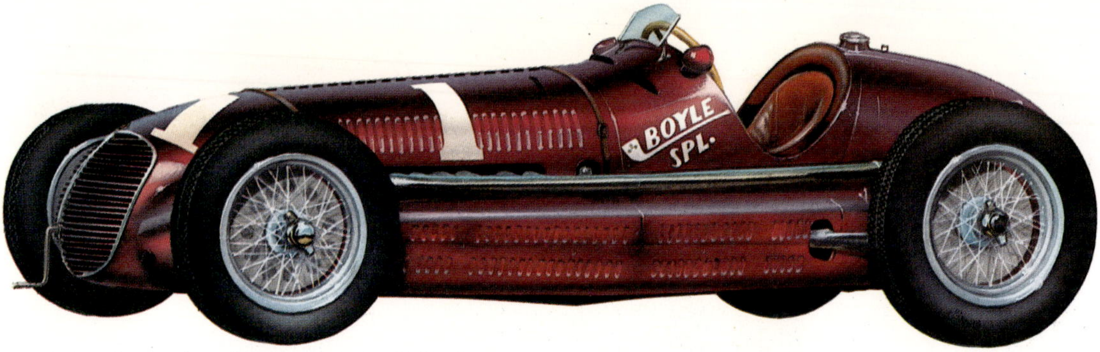

which time the Miller-based Offenhauser engines used as standard by the American track community were being installed in 'roadster' cars, based palpably upon the Maserati which had brought such revolution in 1939.

Maserati 4CLT/48
1948, Italy

In 1938 the surviving Maserati brothers, Ernesto, Bindo and Ettore (Alfieri having died after a bodged appendicitis operation) sold control of their company to industrialist Count Omer Orsi. The factory was moved from Bologna to Modena (although Maserati's badge always retained the trident symbol of the city of Bologna), and with a sound financial base produced many of its four-cylinder voiturette models.

In 1946 revived 4Cs were in the forefront of motor racing's reawakening, and the design was quickly developed with two-stage supercharging and with a tubular chassis frame in place of the former girder type. These developed models were known first as the 4CL, then with the tubular chassis as the 4CLT. At all times the independent front suspension had been by wishbones and torsion bars, with quarter-elliptic leaf springs carrying a rigid axle at the rear. In 1948 the 4CLT appeared in its definitive form with coil-springs for the front suspension tucked inboard out of the airstream and operated by extension lever arms attached to the wheel-hub carriers. This 4CLT/48 model was raced by the quasi-works Scuderia Ambrosiana team and sold well to private entrants.

In their debut two 4CLT/48s were placed 1–2 at San Remo, driven by Alberto Ascari and Luigi Villoresi. The model was known as the San Remo Maserati ever after. Villoresi won four more major races that season, and was second to Wimille's Alfa Romeo in the Italian Grand Prix, beating a new V12 Ferrari in the process. In the winter of 1948–49 4CLT/48s raced in the Argentine Temporada series, a virtual unknown named Juan Manuel Fangio scoring an early international success in one. When Alfa Romeo withdrew for the 1949 season, Maserati briefly assumed their mantle before being overtaken by Ferrari's V12s. Fangio made his name in Europe with a 4CLT, while drivers like Reg Parnell, 'B. Bira' and Dr Farina all won for the trident.

By 1950, however, the claimed 260 bhp of the 78 mm × 78 mm, 1490 cc engine was achieved at the expense of oil-oozing unreliability. The old engine had reached the end of its development and although still competitive in club racing events the Maserati 'fours' were now non-entities at Grand Prix level.

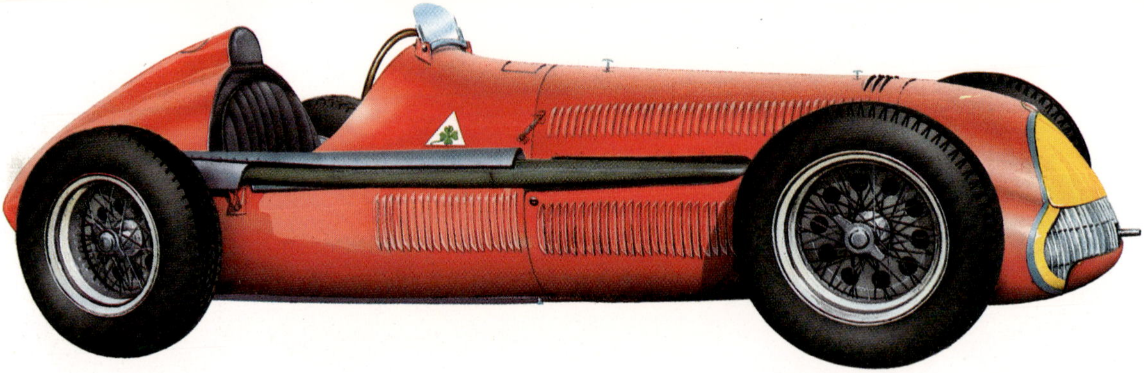

Alfa Romeo 158/159 Alfetta

1947–51, Italy

German industry was shattered by World War II. Auto Union was apparently finished for ever as a motor-racing force and Mercedes-Benz, although its plants were in the Western zone of influence, was in no state to race a bicycle, never mind the most complex Grand Prix cars the world had ever seen. Italy, however, had not suffered such severe destruction, and there is something in the happy-go-lucky Italian character which makes sporting achievement and general enjoyment of life far more vital than any political aim or commercial success.

Consequently the Italian firms of Alfa Romeo and Maserati were well-placed to resume racing the moment peace settled in Europe. In fact Alfa Romeo had continued competition testing and development into 1942 and when the German Army took control in northern Italy in 1943 the seven extant Alfetta 158s were holed-up beyond reach in a remote mountain cheese factory.

In June 1946 two 158s were entered for the St Cloud Grand Prix in Paris. Farina and Jean-Pierre Wimille both led, both retired and established a total team failure unique for Alfa Romeo in this post-war period. The following month saw four 158s at Geneva, two with two-stage supercharging which were placed first in Farina's hands and seventh for Varzi. At Turin Varzi and Wimille were placed 1–2; at Milan Count Trossi, Varzi and mechanic Sanesi were placed 1–2–3.

In 1947 a Grand Prix Formula was recognized allowing $1\frac{1}{2}$ litres unsupercharged. This admitted prewar cars, all that was available, and made a fairer match than the pre-war comparative capacity limitations. Alfa Romeo's 158s entered four events and won all four, normally producing 275 bhp at 7500 rpm while an enlarged low-pressure-supercharger engine gave over 300 bhp for the first time.

Come 1948 and Alfa's season started with the Swiss Grand Prix at Berne, where in practice Varzi was killed in only his second-ever racing accident; but Trossi won the race, and Wimille followed up with victory in the French, Italian and Monza Autodrome Grands Prix to become recognized

◀ *Juan Manuel Fangio of Argentina, five times World Champion Driver, in his works Alfa Romeo 159* Alfetta *at Silverstone, 1951.*

as the outstanding driver of his day. Tragically that winter saw Wimille crash fatally in Argentina at the wheel of a Simca-Gordini and Trossi die of cancer. Alfa's driving team had been gutted, and the company withdrew from racing for 1949, in part due to development time required by postwar production projects, and the need of funds to pay for this work.

In 1950 Alfa Romeo returned in force with their famous 'Three Fs' team of Luigi Fagioli, Giuseppe Farina and the new Argentinian ace, Juan Manuel Fangio. Fangio won at San Remo, Monaco, Spa, Reims, Geneva, and Pescara while Farina was victorious at Silverstone in the British Grand Prix (opening round of the newly-instituted World Drivers' Championship competition), Berne, Bari, Silverstone (for the International Trophy race), and in the Italian Grand Prix at Monza. He narrowly beat his team-mate to clinch the World Drivers' Championship title.

During that year, however, Ferrari's unsupercharged V12 cars, initially in 3.3-litre form and progressively growing towards a full $4\frac{1}{2}$ litres, had pressed Alfa Romeo hard, and for 1951 the 159 model appeared with a staggering 42.6lb supercharger boost and 425bhp at 9300rpm on the test-bench. In this form the cars did about 1.6 miles per gallon and were therefore bloated with extra fuel tankage. Still they had to make one and sometimes two refuelling stops during a Grand Prix distance race, while the slightly slower, unsupercharged cars were capable of running through non-stop, or with one quick top-up at most.

▲ *Works Alfa Romeo team in the woodyard at Berne's Bremgarten circuit, prior to the 1948 Swiss Grand Prix.*

▼ *The fearsome frontal aspect of the Alfetta was a grim warning for drivers of lesser cars when it loomed in their mirrors.*

The cars took 11 first places during the new season, but at Silverstone Gonzalez beat them at last and shattered Alfa's utter supremacy with his unsupercharged Ferrari 375 V12. Fangio became World Champion, however, and Alfa Romeo retired from the Grand Prix scene they had graced for so long. They returned eventually in 1979.

Talbot-Lago Type 26C
1948, France

The old and honoured name of Talbot was sustained by Italian-born Major Antoine 'Tony' Lago who joined the dying Sunbeam-Talbot-Darracq combine in 1933. He managed the famous Talbot-Darracq works at Suresnes and when the combine collapsed in 1935 he found backing to establish an independent Automobiles Talbot company. Looking at the firm's '23CV' six-cylinder production engine, Lago considered it had competition potential and engineer Walter Becchia applied to it an overhead-valve cylinder head with hemi-spherical combustion-chambers. The resulting 160 bhp four-litre engine brought Talbot back into sports-car racing in 1936. In 1937 the team won four of seven races including the second French Grand Prix for sports cars. In 1938 the French race reverted to true Grand-Prix-car status, and a single stripped sports Talbot was the only other finisher behind the Mercedes-Benz flotilla–48 miles behind. . .

For 1939 Suresnes built offset-single-seater $4\frac{1}{2}$-litre Grand Prix cars plus a brand-new central seat *monoplace* intended for a new three-litre supercharged V16 engine designed by Becchia. Lago never found finance to build this unit so the normal six-cylinder was used instead. Raymond Mays retired the single-seater from the 1939 French Grand Prix when its fuel tank split. Just three days after peace with Japan ended World War II, Raymond Sommer drove the Talbot-Lago *monoplace* to second place in a race at the Bois de Boulogne.

In 1946 Louis Chiron drove the ageing car to some fine places and won the revived French Grand Prix with it in 1947. Louis Rosier gave a new Talbot-Lago its debut in the 1948 Monaco Grand Prix, the engine now featuring two high camshafts and producing some 240 bhp at 4700 rpm. A Wilson pre-selector gearbox was used (as in the ERAs) and in 1949 the 9 mpg economy of the powder-blue cars told against the 3–4 mpg of the supercharged $1\frac{1}{2}$-litre Italian machines. Rosier won the Belgian Grand Prix, Chiron the French, and Etancelin, Guy Mairesse and Sommer three other big events. In 1950 Rosier won the Dutch Grand Prix and at Albi, and with wings and lights, won the Le Mans 24-Hours for sports cars. Fangio won the Rafaela 500 in one of these cars and in 1951 Rosier won the Dutch Grand Prix again. Lago's 'trucks' are still racing today.

Ferrari 125
1949, Italy

In 1938 Alfa Romeo withdrew Scuderia Ferrari's sanction to run their racing operations from their remote base at Modena, and brought their racing activities back home to Portello, Milan. The yellow shield and black prancing Horse badge of Ferrari was removed from the cars, and replaced by the green *quadrifoglio* (four-leaf clover) on a white triangle to identify the new works Alfa team.

Enzo Ferrari was taken on to direct operations, but he disliked close control from the Alfa Romeo management and disputes reached a head in his resignation. Part of his severance agreement was that he should not enter racing with any car other than an Alfa Romeo, and when his new Auto Avio Costruzione company did build a couple of 1.5-litre straight-eight sports cars for the 1948 Mille Miglia they were entered simply as '815s' with no marque designation whatsoever. One was driven by Alberto Ascari, son of the late Antonio Ascari of Alfa Romeo P2 fame, and when Ferrari began building cars under his own name in 1946–47 Ascari was again enlisted to drive them.

Ferrari used Alfetta draughtsman Gioacchino Colombo to design his first cars, choosing the power potential and low internal stress offered by a V12 engine layout. Unfortunately the first Tipo 125 V12 was not as powerful as had been hoped, so it was installed in a tiny, abbreviated little chassis which proved to handle diabolically badly. Then, when Alfa Romeo withdrew for the season of 1949 and the Ferrari 125 chassis was lengthened

to tame its handling, Ferrari soared to fame.

The ubiquitous Farina had scored Ferrari's first Formula 1 victory with a 125 at Lago di Garda in 1948 and during 1949 Alberto Ascari developed his driving under the tutelage of team-mate Luigi Villoresi (brother of the late Emilio of Alfa 158 fame); Ascari won the Swiss and Italian Grands Prix while 'Gigi' (as Villoresi was known) won the Dutch event. In 1950 Ferrari turned his attention to unsupercharged engines, designed by his new chief engineer Ing. Aurelio Lampredi.

BRM V16

1950, Great Britain

Just before World War II ERA Ltd had been torn apart by a gentlemanly difference of opinion between Raymond Mays and his technical partner Peter Berthon on the one hand, and the financial backer Humphrey Cook. Mays and Berthon had gone off to launch a scheme to put Britain into Grand Prix racing with a car backed by the best firms in the motoring business.

War delayed the project, but in 1945 Mays gained industrial backing, notably from Alfred Owen and Oliver Lucas. Thus British Racing Motors was formed, to develop an ambitious V16 1.5-litre GP car with Rolls-Royce centrifugal supercharger.

After innumerable delays and problems in austere post-war Britain, the first car was unveiled in December 1949. The V16 engine had a shallow 135-degree vee angle, displaced 1496 cc and could produce a stunning 450 bhp at 10,000 rpm or more. But BRM's project floundered through bad management, overconfidence and under-finance; when it made its over-publicized debut at Silverstone in 1950 and broke on the grid, bitterly misled British enthusiasts jeered and tossed coins into its cockpit. Reg Parnell drove one car to a club race victory at Goodwood that season and in 1951 two cars were placed fifth and seventh in the British Grand Prix, driven by Parnell and Peter Walker. But BRM could not sustain entries in Grand Prix races and when Alfa Romeo withdrew at the end of that season there were no reliable 1.5-litre cars remaining. For 1952–53 Grands Prix were run for two-litre unsupercharged Formula 2 cars, and the V16 became reliable too late.

Cooper-Norton '500' F3

1950, Great Britain

Immediately after World War II a tremendously dynamic movement began in Britain to promote a formula for 'poor man's motor racing'. It all revolved round the idea of building sketchy little four-wheel chassis carrying 500 cc motorcycle engines amidships, behind the driver, so that a direct chain drive to the rear axle could be retained, exactly as on the parent motor-cycle. The most popular engine available was the single-cylinder or 'one-lunger' Norton 500, and from its earliest origins in the Bristol area this form of racing spread vigorously; it crossed the Channel into Continental Europe, reached the USA, and eventually became recognized internationally as Formula 3.

Colin Strang and Clive Lones in Britain were two father-figures of this type of racing and in 1946 Charles and John Cooper, father and son, began copying their pioneering efforts by building 500s from chopped Fiat 500 chassis frames powered by Speedway JAP engines. After minor teething troubles the prototype 550 lb Cooper-JAPs proved most successful. In 1948 a batch of Cooper Mark 2s were offered for sale, with early customers and race winners including Stirling Moss, Peter Collins and Stuart Lewis-Evans. One-litre Cooper hybrids were offered with 988 cc HRD or 996 cc JAP engines while the Norton became standard in Formula 3 trim; in the years 1951–54 Formula 3 Coopers won 64 out of 78 major international races for the class. The Mark 6 model of 1952 used a tubular chassis frame in place of the original box-section affair and in 1955 the Mark 9 adopted disc brakes.

Essentially the mid-engined, transverse leaf-spring all-independent suspension theme was continued unchanged through these years, and in establishing their dominance the Cooper Car Company of Surbiton, Surrey, virtually killed competition within the Formula. When a 1.5-litre Formula 2 class was promulgated for 1957, Charles and John Cooper 'inflated' their mid-engined layout, with its excellent weight distribution and nimble handling, to accommodate a four-cylinder Coventry Climax car engine, and the writing was on the wall for the classical front-engined racing car.

Ferrari 500

1952, Italy

While the number '500' in a Formula 3 Cooper's title indicates that it had an engine of only 500 cc, the figures in this Ferrari's title form part of the Modenese company's unusual car classification system, used for many years. Ferrari classified their cars by the number of cubic centimetres of one engine cylinder. Thus the Ferrari 500 of 1952–53 was a two-litre four-cylinder Formula 2 car, with each cylinder displacing 500 cc: 500 cc × 4 = 2000 cc. Similarly the earlier 1.5-litre supercharged V12 Tipo 125 car had had a unit cylinder displacement of 125 cc: 125 cc × 12 = 1500 cc.

Aurelio Lampredi had developed the simple four-cylinder twin-overhead-camshaft two-litre engine for Formula 2 racing in 1950 and it was installed in a conventional twin-tube chassis frame with transverse leaf-spring, independent front suspension and a De Dion axle layout at the rear. When 1.5-litre supercharged Grand Prix competition died at the end of 1951, organizers were allowed freedom

by the International governing body – the Commission Sportive Internationale – to throw their World Championship Grand Prix races open to unblown two-litre Formula 2 cars. Ferrari, wily as ever, had a well-proven design ready to go.

With the 90 mm × 78 mm, 1980 cc, 170 bhp Ferrari 500, Alberto Ascari proved virtually unbeatable in these two seasons, taking two successive Drivers' World Championship titles. In 1952 he won every Grand Prix he started, six of them, and in 1953 he won five times in eight starts. The only major Grand Prix he did not contest in 1952 fell to team-mate Piero Taruffi in a sister Ferrari 500; and in 1953 newcomer Mike Hawthorn won the French Grand Prix and Dr Farina the German event for the Prancing Horse team. It took Fangio to break Ferrari's stranglehold by winning the last two-litre Grand Prix, at Monza at the end of the 1953 season.

Add to this magnificent success story non-Championship race wins at Syracuse, Marseilles, Saint-Gaudens, La Baule, Pau and Bordeaux, and you can appreciate how the news of another 'Ascari, Ferrari' victory became repetitious in those two remarkable years.

Maserati A6GCM

1953, Italy

When two-litre unsupercharged Formula 2 cars became acceptable for Grands Prix when the Formula was changed for 1952–3, Maserati's spearhead was a development of the sports-racing A6G six-cylinder engine first produced in 1947. Gioacchino Colombo reworked it, and for 1952 it gave 165 bhp from 1988 cc. Maserati often ran Ferrari close – the more so in 1953, with handsome new bodies and the driving of Argentinians Juan-Manuel Fangio and José Froilan Gonzalez. Success came finally at Monza, where Fangio won Maserati their first Italian GP victory.

This was an emotional moment, and it was from the A6GCM car – often erroneously referred to as the 'A6SSG' – that the archetypal front-engined Grand Prix car of the nineteen-fifties, the Maserati 250F 2.5-litre six-cylinder, was subsequently developed for competition in 1954–60. The initials 'GCM' indicated 'Ghisa' – iron (cylinder block – which by this time was light alloy!); 'Corsa' – racing; 'Monoposto' – single-seater.

The Cooper-Bristol took the great Mike Hawthorn to his first major successes.

Cooper-Bristol T23 Mk 2

1953, Great Britain

While Cooper Cars built mainly mid-engined Formula 3, Formule Libre and sports cars well into the mid-fifties, the advent of unsupercharged two-litre Formula 2 racing gave Charles and John Cooper the idea of moving up the motor-racing ladder. The BMW 328-derived Bristol six-cylinder engine was adequately powerful for most competition purposes, was readily available and also very reliable at this time, so the Coopers father and son decided to build a car round it for international Formula 2 racing.

Since the six-cylinder engine was quite long it was considered impractical to install it in the mid-car position behind the driver, and so a conventional layout was adopted, with the engine up-front. The 66 mm × 96 mm, 1971 cc, pushrod overhead-valve engine delivered about 127 bhp and 5800 rpm, and in Cooper's tough box-section chassis frame with independent suspension front and rear by wishbones and transverse leaf-springs, it performed adequately.

Three cars made the type's debut at Goodwood on Easter Monday, 1952, the hitherto virtually unknown Mike Hawthorn bursting to stardom by winning two races in one of the cars, while Alan Brown won a third.

Through the rest of the season Hawthorn went from strength to strength in his Cooper-Bristol and won himself a Ferrari works drive in 1953. He was the first Englishman to drive for a top Continental works team since Dick Seaman, who drove for Mercedes-Benz in 1937–39.

For 1953 a Mark 2 Cooper-Bristol was built around a tubular chassis frame and again proved very popular with British private entrants including Ken Wharton, Alan Brown, and Ecurie Ecosse – the Scottish team for whom Jimmy Stewart drove regularly. Jimmy had a kid brother at home more interested in clay-pigeon shooting than in motor racing; his name was Jackie . . .

Kurtis KK500A

1953, United States

Colorado-born Frank Kurtis began building competition cars in the early thirties, and his company, Kurtis Kraft, was firmly established in 1938, building technically innovative single-seaters for US midget-class and National Championship track racing in large numbers, including Indy '500' winners of 1950–51 and 1953–55. He built the Cummins Diesel 1952 Indy qualifying car, and the legendarily successful centrifugally supercharged Novi V8s. From 1949 some ugly sports cars also emerged. Over 2000 Kurtis Kraft chassis were built, the last front-engined roadster in the early sixties.

In 1953 and 1954 Bill Vukovich won the Indy '500' in the 'Fuel Injection Special' owned by Howard Keck, qualifying at average speeds of 138.392 and 138.488 mph respectively! Jack McGrath, in another KK500 chassis, qualified at 141.033 mph in 1954 to take pole position. Poised for the hat-trick, 'Vukie' was killed in the 1955 '500', involved in another driver's accident.

Mercedes-Benz W196

1954, Germany

Mercedes-Benz returned to racing after World War II as early as 1952; in that year they took three of their 1939 W154/163 three-litre V12 Grand Prix cars to Argentina but were humbled by Froilan Gonzalez handling a modern two-litre Ferrari on a very tight circuit. At Le Mans, Frenchman Pierre Bouillon, driving a Talbot as 'Levegh', tried to cover the 24-Hours single-handed and when in his fatigue he made a mistake with 30 minutes to go, Hermann Lang and Fritz Reiss won the race for Mercedes in a 300SL coupe.

A new 2.5-litre unsupercharged Grand Prix Formula was to begin in 1954 and, spurred on by their Le Mans success, Mercedes-Benz – as usual – spared no expense in preparing for it. They developed their W196 model to employ a brand-new straight-eight engine with desmodromic (mechanically closed) valve gear and Bosch fuel injection; the whole unit was slanted over steeply to one side in order to allow minimal bonnet height. The engine was mounted in a multi-tubular spaceframe chassis with torsion-bar suspension, and was mated to a five-speed gearbox in unit with the rear axle. The M196 engine (Mercedes-Benz used 'W' for 'Wagen' to indicate the car overall, and 'M' for 'Motor' to indicate the engine alone) had dimensions of 76 mm × 68.8 mm, giving 2.49 litres, and it developed around 270 bhp. The German engineers also adopted all-enveloping fully streamlined bodywork with the driver seated centrally in a spacious cockpit.

The cars were late in appearing, finally making their debut at Reims for the French Grand Prix in mid-summer 1954. They won, of course, with Fangio at the wheel of the winning car and Karl Kling's sister W196 second. In three hours or so Mercedes-Benz had again set new performance standards in Grand Prix racing.

At Silverstone the poorly marked turns troubled the drivers of the streamlined cars; for the German Grand Prix a conventional slipper body was adopted and Fangio won. Fangio also won the Swiss and Italian Grands Prix and, with his early-season successes for Maserati being taken into account, was World Champion. During 1955 Mercedes-Benz went from strength to strength, using Moss alongside Fangio, plus Kling, Hans Herrmann and Piero Taruffi as drivers. In the British Grand Prix at Aintree Moss led a Mercedes W196 1–2–3–4 finish, and at the end of the year the team withdrew – having proved their superior skills once more.

◄ *The slipper-bodied exposed-wheel W196 swept all before it through the greater part of the 1954 and 1955 Championship seasons.*

Connaught B-Type

1955, Great Britain

The non-Championship Syracuse Grand Prix, held in Sicily late in 1955, formed an historic landmark in motor racing. The amateur driver Tony Brooks – a dentist by profession – took on the might of the works Maserati team in his under-financed British Connaught and beat them on their native soil. It was the first time that a premier-Formula Continental Grand Prix of any kind had been won by a British driver at the wheel of a British car since Segrave's success for Sunbeam at San Sebastian in 1924.

Rodney Clarke and Kenneth McAlpine – of the famous building family – had founded Connaught Engineering in 1947 to build a sports car for McAlpine to race. They used a 1767 cc Lea-Francis engine and the car went so well that customers queued up for replicas.

Engineer Mike Oliver was one customer, and in 1950 he assisted in building a prototype Formula 2 car using the 'Leaf' engine in modified form. For 1951 nine Formula 2 Connaught A-Types were constructed; the engine was now developed virtually beyond recognition from its Lea-Francis parentage, displacing 1964 cc from 79 mm × 100 mm, and producing 140 bhp at 6000 rpm. The Connaught chassis were excellent, with good balance and handling, but they never had the power to achieve dramatic results.

Then for the new 2.5-litre Grand Prix Formula of 1954 Connaught adopted the Alta four-cylinder engine and clad it initially in a strikingly advanced all-enveloping streamlined body, enclosing the wheels and with a central, single seat. This body was quickly discarded for practical rather than technical reasons and replaced by a normal slipper body, and it was the Connaught in this form that took Brooks to his historic success at Syracuse.

The B-Type 2.5-litre Connaught was known as the 'Syracuse' model thereafter, but industrial sponsorship in sufficient sums never came the company's way and despite some really very enterprising engineering and futuristic schemes which later proved quite realistic and workable, Connaught floundered from one financial crisis to the next and finally collapsed in 1957.

Lancia D50 V8

1955, Italy

Ex-Alfa Romeo engineer Vittorio Jano designed for Lancia a long line of very advanced production cars which handled supremely well; then, for the new 2.5-litre Formula of 1954, the company authorized him to produce a team of full-blown Grand Prix cars. His design was as original as the Mercedes-Benz W196 but had little in common with it.

Jano laid out a V8-cylinder 2.5-litre engine of 76 mm × 68.5 mm and alternatively 73.6 mm × 73.1 mm, finally opting for the longer-stroke unit. Each block of four cylinders had twin overhead camshafts, and four double-choke carburettors provided one inlet to each cylinder. Jano mounted this unit in torsion within his lightweight multi-tubular spaceframe chassis in order to add stiffness to the structure; to obtain minimum handling change during a race as full fuel tanks became empty, and to concentrate the car's essential mass well within its wheelbase, he outrigged the tankage in sponsons on each side.

With as much as 295 bhp at 8500 rpm claimed for this Lancia V8 engine and a very small and light overall package, much was expected from the D50 model; and with the driving team of Alberto Ascari, Luigi Villoresi and Eugenio Castellotti it proved immensely fast. The car handled tremendously well; but once it had been pushed beyond its limit the evenly balanced front and rear ends tended to lose adhesion all at once, and almost before the driver knew he was close to the borderline he would find himself spinning wildly into a ditch. Ascari shone at Barcelona on the car's debut at the end of 1954 and won at Turin and Pau in 1955, but

at Monaco he crashed luridly into the harbour after losing control at the quayside chicane, having inherited the race lead at that moment. He was pulled out of the water unhurt, but within the week crashed fatally while trying his hand in a sports Ferrari at Monza.

This grievous loss was compounded by financial problems at Lancia, and late in 1955 Gianni Lancia ceded the whole of his racing department to the beleaguered Ferrari concern. Ferrari ran much-developed versions of the D50 with more predictable handling through 1956 and 1957, before replacing it with the Jano-designed V6 Dino in 1958. In 1956 Fangio and Lancia-Ferrari won the World Championship titles.

Maserati 250F/1
1957, Italy

Gioacchino Colombo and Alberto Massimino were responsible for the immortal 2.5-litre six-cylinder Maserati Grand Prix car which proved so widely popular in the 1950s Formula. Maserati at Modena went into virtual mass-production of these cars, over 30 being constructed, all eventually sold into private hands. Characteristically, Maserati had always supplied privateers with cars, often before their own works team vehicles were completed, and this was so with the 250F.

Introduced in 1954, the model won its first race, the Argentine Grand Prix, driven of course by the great Juan Manuel Fangio. Later that year and throughout 1955 the cars were generally outclassed by Mercedes-Benz in Grand Prix racing, but they were generally next best, and they utterly excelled in the many minor non-Championship Formula 1 events run in Europe at that time.

The 250F used a multi-tubular spaceframe chassis of the current vogue with double-wishbone and coil-spring front suspension and a De Dion rear end with transverse leaf-spring. A neat gearbox was in unit with the final-drive assembly between the rear wheels and the 84 mm × 75 mm 2.49-litre six-cylinder engine with twin overhead camshafts and three twin-choke carburettors produced around 260 bhp.

In 1956 Moss drove for Maserati, having campaigned a private 250F to such effect in 1954 that he had won his Mercedes-Benz drive for '55. Moss won at Monaco in 1956 but Fangio was too good in the Lancia-Ferrari and won the Championship. Then in 1957 Fangio rejoined Maserati – his second love after Alfa Romeo – and in a classical season won the Argentine, Monaco, French and a tremendously dramatic German Grand Prix to clinch his fifth and final World Drivers' Championship.

During that year a 68.5 mm × 56 mm 2449 cc V12 engine was tried in 250F chassis but follies in sports car racing cost Maserati dear and they were forced to pull in their horns for 1958. Some lightweight 'Piccolo' Maserati 250Fs were built, in one of which Fangio ran his last race – the French Grand Prix of 1958 – and 250Fs in private hands raced on even into the very last 2.5-litre Grand Prix, the US event in 1960.

Vanwall VW1-VW10

1957, Great Britain

From the British viewpoint, the 1957 British Grand Prix was the most significant event of postwar motor racing, for Stirling Moss and Tony Brooks gave Vanwall its first Grande Epreuve victory, and Britain her first-ever home win. Thus started over 20 years of British domination in Grand Prix racing.

Tony Vandervell, industrialist head of the Vandervell Products bearings concern, was the man behind Vanwall. He had been involved with the BRM V16 project in its early days but characteristically became impatient with the dilly-dallying and time-wasting he saw there.

Consequently he bought a series of Ferrari unsupercharged V12 Grand Prix cars which his men progressively modified and fiddled with to form the series of 'ThinWall Special' Ferraris in which drivers like Mike Hawthorn, Nino Farina and Peter Collins charged to several Formule Libre race victories.

For 2.5-litre racing in 1954, Vandervell hired Cooper to build him a simple front-engined chassis, into which was fitted an entirely new two-litre four-cylinder twin-overhead-camshaft engine developed very largely from Norton racing-motorcycle experience – Vandervell being a director of that company. He called the car the 'Vanwall Special'. In 1955 the ephemeral 'Special' tag was dropped; the car became 'Vanwall' pure and simple, and by this time had a full 2.5-litre power unit.

It was obvious that the Cooper chassis was not up to much, and for 1956 Vandervell hired the best people to modify the design. He used Colin Chapman of the burgeoning Lotus Company to design a very light and efficient, true-spaceframe multi-tubular chassis, and Chapman's colleague, aerodynamicist Frank Costin, produced the unusual teardrop body seen here. Moss drove the new Vanwall to win a minor Formula 1 race at Silverstone first time out but was not available for the rest of the season since he was contracted to Maserati. Harry Schell shone at Reims, the Vanwall running among the Ferrari fleet for lap after lap and giving them a very nasty fright indeed.

Vandervell was not afraid of going abroad for good components, and his very powerful four-cylinder 96 mm × 86 mm, 2490 cc cars used German Bosch fuel-injection, Goodyear US-patent disc brakes and so on.

For 1957 Vandervell signed on Stirling Moss, the almost equally talented though less easily motivated Tony Brooks and the very promising but frail Stuart Lewis-Evans to drive his 280 bhp cars, and they came into their own with victory at Liverpool's Aintree circuit. Moss went on to win the Pescara Grand Prix and then the Italian Grand Prix. This gave Vandervell immense personal satisfaction – it had always been his avowed ambition 'to beat those bloody red cars' . . .

In 1958 Vanwall were far and away the top team and won six of the ten Championship-qualifying races – Moss winning the Dutch, Portuguese and Moroccan Grands Prix and Brooks the Belgian, German and Italian. Lewis-Evans was every bit as fast as his illustrious team-mates but lacked the stamina for very long races, and tragically sustained fatal burns when his engine blew up at Casablanca in the final race of the year.

There the Championship had to be decided between Moss and Hawthorn, Vanwall and Ferrari, and while the Drivers' title went to Hawthorn, Vanwall achieved the World

Constructors' Championship.

Vandervell had proved his cars, his British cars, the best in the racing world; but with his health failing, and deeply affected by Lewis-Evans' death, he withdrew from competition thereafter – only one or two half-hearted entries following from Vanwall until they dried up completely in 1961.

▲ *Tony Brooks at Silverstone in a 1958 Vanwall showing off its glorious lines and scientifically designed air intake and exit systems.*

◄ *Frank Costin's glorious high-tailed teardrop body on the Vanwall played a major part in the car's 1957–58 success story.*

Porsche Type 804
1962, Germany

After many years' participation in sports car events, Porsche found their way into Formula 2 racing during the seasons of 1957–58, using central-seat versions of their 1.5-litre flat-four-cylinder RSK sports-racers. For 1959 the Stuttgart-Untertürkheim company produced 'proper' Formula 2 cars with single-seat slipper bodies and torsion-bar suspension; having raced successfully throughout 1960 they were well-prepared for the new Formula 1 introduced in 1961 – effectively the familiar 1.5-litre 1957–60 Formula 2 regulations, with extra weight and safety restrictions.

During 1960 Dan Gurney and Jo Bonnier drove these Porsche Type 718 cars regularly and they proved quite competitive, their air-cooled 85 mm × 66 mm engines delivering about 155 bhp at 7500 rpm. But they had neither the power nor the handling actually to win anything, and for 1962 a flat-eight-cylinder air-cooled Type 804 model replaced these bulky 'Old Ladies' of the Grand Prix world.

The Type 804's beautifully made engine, with a massive cooling fan perched above it, had bore and stroke dimensions of 66 mm and 54.5 mm respectively to provide 1498 cc, and it delivered a claimed 185 bhp at 9200 rpm. Dan Gurney shone in the new silver cars even though, in truth, the Porsche had the power and handling characteristics of a road-going VW Beetle. With a little luck he won the 1962 French Grand Prix at Rouen for the German team; he dominated the German Grand Prix at Nürburgring in terms of sheer pace but when it rained could not improve on third place. On Porsche's doorstep in the non-Championship Solitude Grand Prix just outside Stuttgart, Dan and Jo Bonnier sent 350,000 German spectators wild with delight by finishing first and second in the teeth of strong British opposition.

But Porsche found it hard to justify the budgetary outlay involved in making racing cars unlike those they produced for sale, and at the end of the 1962 season they withdrew. Despite immensely expensive sports-car racing projects in succeeding years, Porsche have, at the time of writing, yet to return to the white heat of Grand Prix competition.

Jo Bonnier in the works flat-eight air-cooled Porsche 804 during the 1962 British Grand Prix at Aintree.

BRM P57 V8

1962, Germany

After the disasters of the V16 BRM, the original BRM Trust sold out the company complete to Alfred Owen's private industrial combine, the Owen Organization, who developed the V16 to competitive form for minor Formule Libre events and during the later fifties ran a new 2.5-litre four-cylinder P25 Grand Prix design in World Championship events. Yet again enthusiasm exceeded practicality and success was limited until 1959 when Jo Bonnier won the Dutch Grand Prix in one of the cars. By the end of 1961 BRM did not have another victory in anything significant, and Sir Alfred Owen's brother Ernest and the other members of the family board were making noises about the small fortune being absorbed by BRM's continued operation.

Sir Alfred made it clear to BRM's personnel that 1962 was a make-or-break season. They were to win something substantial, or the team operation would be folded up. Peter Berthon had designed a 1.5-litre V8 engine known as the P56 for the new Formula, and this was subsequently developed by ex-Rolls-Royce engineer Tony Rudd and mounted in a spaceframe chassis and body of his design, known as the P57. It appeared in practice for the 1961 Italian Grand Prix and impressed as one of the smallest and prettiest Formula 1 cars ever seen.

Early in 1962 the car won the first heat of its debut race, the Brussels Grand Prix, driven by Graham Hill. It went on in Hill's hands to win at Goodwood and then again in an epic near-dead-heat with Jim Clark's new Climax V8-powered Lotus 24 at Silverstone. The World Championship season opened at Zandvoort for the Dutch Grand Prix and Graham won in magnificent style. Hill fought a season-long duel with Clark and Lotus for the Championship titles; he and BRM triumphed, with victories in the German, Italian and South African Grands Prix. In 1963 Hill won at Monaco and in the United States Grand Prix, and team-mate Richie Ginther excelled once more.

The V8 spaceframe BRMs continued to provide private entrant service into 1965 with their 68.5 mm × 50.8 mm, 1497.7 cc engines giving 188–202 bhp at 11000 rpm. These engines also powered other chassis by Lotus, Brabham, BRP and Scirocco-Powell.

Graham Hill at Goodwood in 1962 demonstrating why the new V8 BRM was nicknamed the 'Stackpipe' at that time.

Lotus-Climax 33

1964, Great Britain

Colin Chapman put his Lotus Engineering company in the forefront of advanced motor racing during the fifties with a series of very lightweight and aerodynamically efficient sports-racing cars. He was expert in building multi-tubular spaceframe chassis of which every single frame component was put to work in either tension or compression. It was this expertise that won him the Vanwall consultancy contract during 1955, and in 1956 he introduced his own prototype single-seater Lotus, a front-engined device aimed at Formula 2 and using a Coventry Climax FPF four-cylinder engine.

In 1960 he took to the mid-engined configuration with the Lotus Type 18, and its performance set the standards of the day, shaking even Cooper's grasp on Formula 1 in non-Championship events, if not in World Championship rounds. Stirling Moss won the Monaco Grand Prix that season in Rob Walker's private Lotus-Climax 18, and won the Monaco and German Grands Prix in a similar Walker car in 1961. At the end of that year Innes Ireland gave Team Lotus their first Grande Epreuve victory with the much-improved and very sleek Lotus 21 spaceframe chassis car; for 1962 fast-developing Jimmy Clark became the Lotus number one driver, and Chapman devised a new car for him.

There had been constant problems with tubular chassis chafing and puncturing the aluminium fuel tanks strapped to them, and the Lotus design team was at this time developing a backbone girder chassis for a new road car–which eventually emerged as the Elan.

Tests showed this slender backbone to be immensely rigid, and Colin visualized widening it until there was room for a driver to sit between its side sections, and boxing in those side sections to create fuel tank space. Thus the epochal Lotus 25 was evolved, but it was a new idea and to hedge his bets Chapman also laid down a spaceframe alternative to pick up the latest all-independent suspension system developed for the Lotus 25.

The Lotus 25 chassis was formed in aluminium sheet, fitted over internal sheet bulkheads with not a tube in sight. It looked like an open bathtub in which the driver sat with fuel on each side of him. The new Coventry Climax FWMV V8 engine was behind his shoulders. The 63 mm × 60 mm engine had a displacement of 1494 cc, and produced about 190 bhp at 8500 rpm. Climax engines, incidentally, were now used by most British chassis-constructors, including Brabham, Cooper and Lola; only BRM, Porsche and Ferrari still made not only their own racing car chassis, but also the engines and transmissions.

The Lotus 25 made its debut in the 1962 Dutch Grand Prix and Clark led until suffering transmission troubles. He first won a Grande Epreuve in the car at Spa in Belgium, then won the British and United States Grands Prix. He led most of the way during the South African Grand Prix, but lost the World Championship when his engine lost its oil.

The Type 25 grew up into the wide-tyred 200-horsepower Lotus 33 as driven to victory here at Silverstone in 1965 by Jim Clark.

Jimmy Clark was virtually unbeatable during 1963 with the Lotus 25, the world's first successful monocoque-chassised Grand Prix car, originated as seen here in 1962.

During an incredible 1963 season, however, the Lotus 25 theme was developed and Jimmy Clark won a record seven of the year's ten Championship rounds to claim his and Lotus's first world titles. In non-Championship races, Clark and the 25 won five more events.

The Lotus 25 was the tap-root of modern racing car design and its immediate derivatives included the Indianapolis Lotus-Ford 29 and 38. These cars turned the Indy establishment away from the Offenhauser roadsters, reminiscent of 1939 Maseratis, towards the pure performance potential of road-racing-derived modern chassis.

In 1964 the much-improved, big-wheeled Type 33 was introduced by Lotus and with updated 25s saw Clark challenging hard for the Championship yet again, only losing it on the last lap of the last race – in Mexico City. In 1965 Clark took the Championship convincingly in his Lotus-Climax 33s, winning six races – five consecutively. He also won the Indy 500 in a Lotus-Ford 38 with 4.2-litre V8 engine.

As the Lotus 33s with two-litre engines continued racing in the first two seasons of the new three-litre Grand Prix Formula in 1966–67, so all constructors save Brabham followed the Lotus monocoque lead. Colin Chapman's innovation had succeeded in 1962 where Gabriel Voisin's had failed in 1923.

Jack Brabham drove his BT19 to victory in the French Grand Prix – and then on to the Drivers' World Championship. The engine was developed from an Oldsmobile original.

Repco-Brabham BT19
1966, Australia

Coventry Climax announced during 1964–65 that they would not be building a new engine for the three-litre Grand Prix Formula proposed for 1966. This move left such British specialist constructors as Lotus, Brabham and Cooper out on a limb. Lotus had Cosworth Engineering build them an engine with Ford money; Cooper turned to Maserati in Italy, who wheeled out a three-litre version of their age-old 2.5-litre V12 engine of 1957. Brabham looked to his native Australia and the Repco specialist performance equipment company there, who had helped fund his production racing car company upon its inception in 1961–62.

Repco took a very simple Oldsmobile F85 production V8 engine with an alloy block and developed it extensively for Formula 1 use. It offered a maximum 315 bhp at 7250 rpm from 88.9 mm × 60.325 mm cylinders, displacing 2995.7 cc; there was a single overhead camshaft per bank. Installed in a very light and practical spaceframe chassis designed by Brabham's associate Ron Tauranac, this was to prove sufficient.

The chassis used originally had been laid down for a still-born flat-sixteen-cylinder 1.5-litre Climax engine in 1965, and it was quickly adapted to accept the new Repco engine. Running against H16 BRM, V12 Ferrari, V12 Eagle-Weslake, V12 Maserati and ultimately V8 Cosworth-Ford engines, this unit did amazingly well. Jack Brabham's BT19 won the 1966 French Grand Prix – the first to fall to a driver in a car bearing his own name – and then won the British, Dutch and German Grands Prix consecutively, clinching the World Championship titles for himself and for his cars.

In 1967 Ron Tauranac did 'a proper job' on the chassis, building the very compact BT24 car with which Denny Hulme won the Drivers' title, pipping his team-leader, who was content that his marque was for the second successive year Champion of the World.

Eagle-Weslake
1966, United States

Dan Gurney was an American enthusiast and a very talented Grand Prix driver. He won Grand Prix races for Porsche and Brabham-Climax in the 1.5-litre Formula of 1961–65 and then gained the backing to build his own Eagle cars for the new three-litre Grand Prix Formula of 1966 and for Indianapolis in America, in which race he had first interested Colin Chapman, and so brought about the Lotus-Ford rear-engined revolution there.

Dan and his backers established All American Racers Inc. at Santa Ana, California, late in 1965 and used Len Terry, designer of the Indy-winning Lotus 38, to draw his prototype Eagle cars for '66. Ties were then forged with Harry Weslake Research at Rye in England who were to build a BRM-like three-litre V12 engine for the Formula 1 variant of the Indy monocoque chassis. This wing of the company was named Anglo-American Racers and while development proceeded with the V12 engine Dan sorted out the first T2G Formula 1 Eagle chassis by using a 2.75-litre Climax FPF four-cylinder engine. He scored his first World Championship points for his own marque in the French Grand Prix that year, coming fifth, and was fifth again at Mexico City. The V12 Weslake engine was slow in arriving and proved a fretful baby, but in 1967 Dan became the first American driver in what was effectively an American car since Jimmy Murphy in 1921 to win a Grande Epreuve, when his Eagle-Weslake V12 won the Belgian GP.

Unfortunately niggling problems hit the V12s, and the Formula 1 activities of AAR faltered into 1969 and finally ceased. Dan's backers were primarily interested in US-style track events, and the Indy 500 in particular – which event an Eagle Offenhauser won in 1968; Eagle were to dominate that form of competition for many years.

Still, Daniel Sexton Gurney's heart was in road racing, and even today he remembers that Spa race with immense satisfaction.

Dan Gurney's Anglo-American Eagle V 12

Lotus 56B Turbine

1968, Great Britain

Peugeot, Delage, Mercédès and Ballot had done very well in the 500 Miles Indianapolis track classic in the World War I years, but European interest waned until the success of the Maserati 8CTF, influential in America. During the late forties and the fifties, Indy roadster design largely stagnated, the Miller-derived Offenhauser engine prevailing, although features such as disc brakes and fuel injection were adopted here earlier than in Formula 1.

In 1961 Jack Brabham finished ninth at Indy in a GP-derived 2.75-litre mid-engined Lotus Climax; in 1962 Dan Gurney drove a mid-engined Mickey Thompson car, and took Colin Chapman along; in 1963 the first Lotus-Ford Indy cars were placed second – and in 1964–5 the rear-engined revolution gripped US track racing.

In 1967, Andy Granatelli of the STP Corporation ran a gas-turbine-powered four-wheel-drive car at Indy and very nearly won the race. For the following year he hired Lotus to build a team of turbo-cars, and the dramatic wedge-shaped four-wheel-drive Type 56 was the result. The cars did not have a lot of power, but they handled supremely well. They were heading for an historic success when, with only nine laps of the 200 remaining both Art Pollard and leader Joe Leonard were side-lined by fuel-pump failure. Graham Hill's car had lost a wheel.

For 1969 the American track establishment reacted by effectively banning turbine power and eventually four-wheel drive, while spin-offs from the Lotus 56 programme included a flurry of unsuccessful four-wheel-drive Grand Prix cars and – in 1971 – an uncompetitive Type 56B gas-turbine Formula 1 car.

Matra MS11 V12

1968, France

In the early-sixties there was a tiny French specialist racing-car manufacturer named René Bonnet, which built monocoque Formula Junior cars after the Lotus fashion, but was very short of money. The mighty Matra aerospace company had a glass-fibre and plastics division which produced body panels for Bonnet's single-seater and long-distance sports-racing cars, and when Bonnet ran short of money, Matra's management absorbed his concern with the idea of promoting their engineering expertise through motor racing.

In 1964 a new one-litre Formula 2 for racing engines and one-litre Formula 3 for production-based engines came into effect. Matra built beautifully constructed monocoque single-seater cars for both classes, using British Cosworth-Ford engines. They proved very successful. In 1966 Ken Tyrrell, formerly Cooper Formula 2 team patron, was looking for a new chassis for the young driver he had brought to prominence in Formula 3 during 1964. That driver was Jackie Stewart. Ken was asked by Jean-Luc Lagardère, Matra's dynamic young chief executive, to try one of the French chassis, and Stewart was instantly enthralled.

The Tyrrell Matra Formula 2 team raced very successfully through 1966–67, and in the latter season Ken saw the debut of the new Cosworth-Ford Formula 1 engine. He knew it might be offered to teams other than Lotus in 1968, and so managed to attract Stewart from BRM, and arranged for Matra to build chassis to accept the Cosworth-Ford engine; with sponsorship from the French ELF oil company he formed a Grand Prix team. In 1968 Stewart's Matra MS11-Ford won the Dutch, German and United States Grands Prix while Matra's own driver Jean-Pierre Beltoise handled the MS10 car with the Matra V12 engine installed. In 1969 a new MS80 car for Stewart proved outstanding with Ford power, and Jackie won six Grands Prix to clinch his first World Championship. Matra became World Champion constructor, and the object of their absorption of René Bonnet's failing little company five years previously was achieved.

◄*Jackie Stewart driving his Anglo-French Matra-Cosworth to victory in the 1968 Dutch Grand Prix at Zandvoort on the North Sea coast.*

►*Jacky Ickx sliding his Ferrari 312B2 at Zandvoort during the 1971 Dutch Grand Prix.*

Ferrari 312B Flat 12

1970, Italy

Ferrari pulled themselves out of their 1955 trough by accepting the Lancia D50 gift; by 1959–60 their front-engined cars, though powerful, were old-fashioned, overweight and ill-handling compared to the rampant Britons. When the 1.5-litre Formula began, in 1961, Ferrari was the only team properly prepared with a really powerful multi-cylinder engine – a V6 – and they won the Championships that year and also in 1964. For the new three-litre Formula in 1966 they developed one of their age-old sports-racing V12 engines to suit, and won races occasionally though in many cases unconvincingly into 1968. Then in 1969 their cars lost their competitive edge, and Ferrari fortunes slumped.

It was during this season that chief engineer Mauro Forghieri was developing with Franco Rocchi a new twelve-cylinder engine in which the cylinder banks were laid horizontally, opposing one another on a common crankcase. This flat-twelve engine placed its mass very low in the car, and thus potentially aided the car's cornering abilities; it was to develop into an extremely powerful unit, which survived ten years in Formula 1.

The original 1.5-litre Ferrari flat-twelve had been developed by Forghieri for Formula 1 in 1964–65. A later development was the two-litre 212E engine which won the 1969 European Mountain Championship in a sports Ferrari chassis, and in 79 mm × 52.8 mm, 2998.5 cc form the unit made its three-litre Formula 1 debut in 1970 in South Africa. It was claimed to produce 460 bhp at 11,600 rpm but was thirsty for fuel. In that year's Austrian Grand Prix, Jacky Ickx gave the new 312B (three-litre twelve-cylinder 'Boxer') its first victory, with team-mate Clay Regazzoni second. In Italy Regazzoni won; in Canada Ickx led another 312B 1–2 finish and the result was the same in Mexico City. New 312B2 models were developed for 1971 with revised suspensions, but the 1970 model was outstandingly beautiful in mid-engined Grand Prix car terms and has gone into racing history as Ferrari's Grand Prix turning point.

Ferrari were to have a low time again – in 1973 – but bounced back from 1974, winning three consecutive Championships in 1975–77, then another in 1979.

Lotus-Ford 72

1970–71, Great Britain

From 1968 to 1974 seven successive World Championship Grand Prix titles were won using the Ford-financed Cosworth DFV V8 three-litre engine. Ferrari took the Constructors' titles from 1975 to 1977 but in 1978 the ageing though still very competitive British engine won again. The compact and very practical DFV won more Grand Prix races, well over 100, than any other engine in the history of motor racing and in modified form added victories in the Le Mans 24-Hours sports car endurance race and in the superfast Indianapolis 500-Miles track classic.

Lotus were deeply involved with the inception of this engine. Late in 1965 Colin Chapman asked Keith Duckworth of Cosworth Engineering if he thought he could produce a Grand Prix power unit for the forthcoming three-litre Formula. Duckworth considered that he could, and that development and launching costs would approach £100,000. Chapman interested Ford of Britain in the idea, and with Ford's backing and Ford lettering on the cam-boxes the DFV made its debut in the 1967 Dutch Grand Prix in the stern of a brand-new Lotus Type 49. This car was designed by Colin Chapman and Maurice Phillippe as a no-nonsense chassis which would not cause any development problems and so divert attention from the engine itself during its teething stage. In fact the chassis and engine proved themselves first time out in a fairy-tale debut at Zandvoort. Graham Hill started his car on pole position after setting fastest time in practice, and when Hill retired, team-mate Jimmy Clark then took over to lead to the finish, the cars setting fastest lap – and an entirely new performance standard – on the way.

The Lotus 49 went on to win 11 Grand Prix races and numerous minor Formula 1 events until it was replaced in 1970 by the Type 72 model, which itself created another Grand Prix design revolution. In the old 49 Chapman had created a monocoque forward cockpit nacelle which simply terminated in a vertical bulkhead behind the driver's shoulders. The DFV engine was designed to bolt directly to that bulkhead, and to carry the gearbox and

▶ *Emerson Fittipaldi during the 1973 Spanish Grand Prix, with the Lotus 72.*

◀ *The great Swedish driver Ronnie Peterson in the Lotus 72 during the 1975 French Grand Prix at Ricard-Castellet on the Mediterranean coast.*

rear suspension loads itself, without the aid of any extra chassis structure. This became a standard feature in other DFV-powered cars which appeared during 1968, notably Matra and McLaren.

Now, the Lotus 72 continued this theme – which in fact survives to this day – but its forward monocoque nacelle was a perfect wedge-shape intended to create the maximum down-force from wind effect at speed, and so literally push itself down onto the track surface. This shape had been developed by Lotus in their 1968 Indianapolis Type 56 turbine-powered four-wheel-drive cars. In the Formula 1 Type 72 it was allied with a very sophisticated suspension system, which replaced the hitherto virtually universal coil-spring and wishbone systems with torsion-bar springs front and rear, leaving the space between the suspension links on each side uncluttered and open to the airflow. Brakes were also mounted inboard at front and rear to eliminate heat-soak through the wheel rims into the tyres, and this helped to allow the fitting of very soft and sticky Firestone tyres which on other cars would normally have grown so hot as to melt.

The Lotus 72 was above all a carefully integrated design, and once teething troubles had been sorted out early in 1970 it progressed from strength to strength; it brought Jochen Rindt his World Championship title – unhappily posthumous, since he was killed in a practice accident at Monza. In 1971 a rebuilt Lotus team continued to campaign the Type 72 and in 1972 Emerson Fittipaldi dominated the Formula 1 season with it, winning the World Championship titles. In 1973 Lotus took the Constructors' Championship with the 72 while Jackie Stewart won the Drivers' crown in a Tyrrell. It was the first time since 1958 that the Championships had been split between two teams. Then in 1974 Ronnie Peterson won very different Grands Prix, at Monaco, Dijon and Monza on both very slow and very fast circuits, to prove the old 72's continuing competitiveness. But during 1975 Goodyear's tyre monopoly took effect and Lotus found the special tyres required for the 72 no longer available.

The car abruptly fell from favour, tyre requirements made what had been its great attributes now obsolete, and a truly great Grand Prix car faded into dignified retirement.

Tyrrell-Ford 006/2, 1973.

Tyrrell-Ford 006/2
1973, Great Britain

When Matra ceased allowing Ken Tyrrell to run Ford engines in their chassis in 1970, the English one-time timber-merchant bought chassis instead from the new March Engineering concern. Meanwhile he realized that to provide his driver Jackie Stewart with the very best machinery, he ought to attempt building his own car. To this end, and in utter secrecy, he contracted ex-Ferguson engineer Derek Gardner to build an experimental Formula 1 car to accept the DFV engine. The car was ready in September 1970 and Stewart shone with it in the Canadian and United States Grands Prix to assure the Tyrrell team of sponsorship for their brand-new cars in 1971.

During 1971 Stewart's incredible driving prowess brought Tyrrell victories in the Spanish, Monaco, French, British, German and Canadian Grands Prix, and team-mate Francois Cevert won the US event. Stewart was World Champion Driver for the second time, and Tyrrell-Cosworth became World Champion Constructor in their first full season of competition.

In 1972 Stewart was briefly laid low by an ulcer, and although he won the Argentine and French Grands Prix for Tyrrell the season was dominated by Fittipaldi's Lotuses. Stewart decided that 1973 was to be his last racing season before retirement, and Tyrrell built a new model, numbered Tyrrell 006/2. With this car, short and stubby, highly swerveable and designed as a machine from which only the very finest driver could extract the utmost, Stewart won its debut race at Silverstone, then added victories in the Belgian, Monaco, Dutch and German Grands Prix. With his success at Monte Carlo he equalled the late Jim Clark's record of 25 Grande Epreuve victories, and his Dutch and German victories added two more to set the all-time record of 27 World Championship-qualifying Grand Prix wins.

The Tyrrell 006/2 was driven by new Tyrrell team-members Jody Scheckter and Patrick Depailler in 1974 before joining the World's largest collection of classic racing cars at Donington Park, near Derby, England where it can be seen—occasionally in action—to this day.

McLaren-Ford M23
1976, Great Britain

Bruce McLaren began his Grand Prix career as a youthful team-mate to Jack Brabham in the Cooper team in 1959–60. When Brabham left, Bruce became Cooper's number one and stayed with them until 1965. Meanwhile he had established his own private racing team, first campaigning and latterly building his own Tasman Formula and sports-racing cars from an early start in 1964.

Come 1966 and the start of the three-litre Formula, Bruce built his own Grand Prix cars, and in 1968, when he adopted the Cosworth-Ford DFV engine, the papaya-coloured McLarens began to win Grand Prix races. Bruce was tragically killed while testing a sports car at Goodwood in 1970, but his team survived him and after a period in the doldrums came back to prominence at Indianapolis in 1971 with a startling wedge-shaped projectile based on the Lotus 72 aerodynamic form.

For 1973 a Formula 1 version of this McLaren M16 Indy car was constructed, known with Cosworth engine installed as the

1976 McLaren-Ford M23.

M23. With backing from Yardley Cosmetics McLaren had a good season, with Denny Hulme and Peter Revson winning several Grand Prix races. The M23 cars were repainted to display Marlboro tobacco company livery in 1974 when Emerson Fittipaldi joined the team, and he clinched the World Championship for McLaren with a string of consistent if generally uninspired race places to support a handful of outright wins.

The M23 continued in ever-modified form through 1975, and in 1976 James Hunt joined the Marlboro-McLaren team. Hunt had a torrid season in which he equalled Jim Clark's record of seven Grand Prix victories, was disqualified from two and reinstated in one, finally beating Niki Lauda and Ferrari to the world titles in the final round of the series, at Mount Fuji in Japan.

With two World Championships to its credit, the McLaren M23 joined the select ranks of classic Grand Prix cars. Some still raced as late as 1979 in national F1 events.

James Hunt had a superb run of hard-fought success during the 1976 season in his Marlboro-sponsored McLaren M23s. Here in the Italian Grand Prix at Monza he was 'regulated' out of contention by a dispute over the fuel the team was using, and although this was the third legal wrangle of a troubled season afflicting Hunt and the McLaren team they won through to steal the Championship titles in the final race of the season, at Mt Fuji, Japan.

Lotus-Ford 79

1978, Great Britain

Special aerodynamic aids had appeared in earnest in Grand Prix racing on the superfast Spa circuit in 1968, when Brabham and Ferrari appeared for the Belgian Grand Prix with little wings on struts above their engine bays, and with vanes on the nose to balance them out. The wings were set at such an angle that the airflow over them forced the car down onto the roadway. The effect was to increase tyre grip in cornering, braking and acceleration. Since this aerodynamic effect, which might add 250–400lb to the apparent weight of the car at speed, was not actually extra mass to be accelerated, cornered and decelerated, it was virtually a something-for-nothing development. It certainly contributed mightily to rapidly increasing lap speeds. In 1969 wings had become so huge that they became unstable, and regulations were made to limit them severely. These basic limits survived with little alteration into the mid-seventies by which time wings on racing cars were vital components of the design.

In 1976 Lotus were in deep trouble, trying to find a successful replacement for the much-loved Type 72. Colin Chapman, aerodynamicist Peter Wright and ex-BRM engineer Tony Rudd studied the problems, and considered that use should be made of the air going underneath the car, as well as that which passed over it. They devised a car which effectively carried wings on each side of its centre section; the space beneath the 'wings' was open to the airstream at front and rear, but boxed in on the outside and closed off against the road there by a sliding skirt system.

This car, the Lotus 78, made its debut in 1977 and set new performance standards with staggering speed round corners. Mario Andretti, Lotus number one driver, said, 'It feels like it's painted on the road'. Mechanical troubles in mid-season with development Cosworth DFV engines cost Andretti and Lotus the World Championships, but in 1978 an improved car, now the Lotus 79, joined the 78; Andretti and team-mate Ronnie Peterson achieved total domination, chassis excellence giving the Cosworth-Ford engine another World Championship and beginning a new age – that of 'ground-effect' racing cars.

Renault RS10 Turbo

1979, France

During 1979 several strands of technical advance in Grand Prix racing interwove and created a season of remarkable interest. We had the Lotus 78/79 ground-effects aerodynamic form being developed beyond Lotus's own reach by the French Ligier team, by Brabham (using Alfa Romeo V12 engines) and by Frank Williams' long-unsuccessful team making the best use of Arab sponsorship money. We had Ferrari building the brutal-looking 312T4 model in which their very successful and powerful flat-twelve engine was mounted in a chassis which made perhaps better use of the air going over the body top than any had previously. Their broad flat-twelve engine obstructed airflow through the ground-effects side pods – a problem which did not afflict the Cosworth-Ford-powered brigade with their engine's narrow bottom end, and which Alfa Romeo engineers had side-stepped by sparing little expense to build a V12 engine in place of the flat-twelve which had been used in Formula 1 by Brabham since 1976.

Then we also saw Renault using the 1.5-litre supercharged option in the Formula 1 regulations, with a V6-cylinder engine boosted by an exhaust-driven turbo-supercharger. In this system a small impeller vane is introduced into the flow of exhaust gases, and a spindle couples it to another impeller which forces pressure air back into the induction side of the engine. This turbocharger arrangement had been used successfully at Indianapolis for many years, then in sports-car racing, and Renault had introduced it to Grand Prix competition in mid-1977. By mid-1979, with a ground-effects chassis to make best use of the 550 bhp available from their tiny engine, Renault were competitive and they won the French Grand Prix – Jean-Pierre Jabouille achieving this signal honour for Renault in their first Grande Epreuve success since Szisz's first-ever Grand Prix victory of 1906.

But it was the Patrick Head-designed Saudia-Williams '07 which set the ground-effects car standard for the year, carrying drivers Alan Jones and Clay Regazzoni to several victories – the Swiss in the twilight of an honourable career. The Williams was neat, practical, uncomplicated and efficient – a perfect foil to the pioneering sophistication of the Renault.

Study of the modern ground-effects Grand Prix car unclad: Saudia-Williams at Silverstone, 1979, showing the suction pods either side of the slender monocoque chassis nacelle.

Index

Page numbers in *italics* refer to illustrations.